# Mycorrhizas

**Bibliothèque et Archives nationales du Québec and Library and Archives Canada cataloguing in publication**

Fortin, J. A. (J. André)

Mycorrhizas: the new green revolution

Translation of: Les mycorhizes.
Includes bibliographical references and index.

ISBN 978-2-89544-154-0

1. Mycorrhizas. 2. Mycorrhizas in agriculture. 3. Mycorrhizal fungi. 4. Symbiosis. I. Plenchette, Christian, 1943- . II. Piché, Yves, 1950- . III. Title.

QK604.2.M92F6713 2009 579.5'17852 C2009-941269-1

J. André Fortin
Christian Plenchette
Yves Piché

Translated by Andrew P. Coughlan

# Mycorrhizas

## THE NEW **GREEN REVOLUTION**

ÉDITIONS
MULTIMONDES

ÉDITIONS MULTIMONDES
930, rue Pouliot
Québec (Québec) G1V 3N9
CANADA
Telephone: 418 651-3885
Toll free telephone: 1 800 840-3029
Fax: 418 651-6822
Toll free fax: 1 888 303-5931
multimondes@multim.com
http://www.multim.com

DISTRIBUTION AU CANADA
PROLOGUE INC.
1650, boul. Lionel-Bertrand
Boisbriand (Québec) J7H 1N7
CANADA
Telephone: 450 434-0306
Toll free tel.: 1 800 363-2864
Fax: 450 434-2627
Toll free fax: 1 800 361-8088
prologue@prologue.ca
http://www.prologue.ca

DISTRIBUTION EN FRANCE
LIBRAIRIE DU QUÉBEC/DNM
30, rue Gay-Lussac
75005 Paris
FRANCE
Telephone: 01 43 54 49 02
Fax: 01 43 54 39 15
direction@librairieduquebec.fr
http://www.librairieduquebec.fr

DISTRIBUTION EN BELGIQUE
La SDL Caravelle S.A.
Rue du Pré aux Oies, 303
Bruxelles
BELGIQUE
Telephone: +32 2 240.93.00
Fax: +32 2 216.35.98
Sarah.Olivier@SDLCaravelle.com
http://www.SDLCaravelle.com/

DISTRIBUTION EN SUISSE
SERVIDIS SA
chemin des chalets 7
CH-1279 Chavannes-de-Bogis
SUISSE
Telephone: (021) 803 26 26
Fax: (021) 803 26 29
pgavillet@servidis.ch
http://www.servidis.ch



ISBN 978-2-89544-154-0

Legal deposit – Bibliothèque nationale du Québec, 2009
Legal deposit – National Library of Canada, 2009

*Éditions MultiMondes acknowledge the financial support provided by the government of Canada through the Book Publishing Industry Development Program (BPIDP) for the edition of this volume, and the Canada Council for the Arts for the support accorded to their publication program. We also acknowledge the Société de développement des entreprises culturelles du Québec (SODEC) for their support during the editing and promotion of this book.*

*Government of Quebec – Editors Tax Credit program – managed by SODEC.*

50%

Mixed Sources
Product group from well-managed forests and other controlled sources
www.fsc.org Cert no. SGS-COC-003885
© 1996 Forest Stewardship Council

This book has been printed using vegetable-based inks on acid-free chlorine-free paper containing 50% recycled material, of which 15% is post-consumer waste.

IMPRIMÉ AU CANADA/PRINTED IN CANADA

*To Bernard Bélanger.*
*A visionary entrepreneur who was one of the first*
*to understand the potential of mycorrhizas,*
*and who has succeeded in transforming part*
*of our dreams as scientists, into reality.*

# Acknowledgements

First and foremost, we would like to thank all the students that we have had the pleasure to work with over the years; their studies have taught us most of what we know today. We would also like to thank our colleagues Yolande Dalpé, Chantal Hamel, Susan Parent, Marc St-Arnaud and Martin Trépanier, who read the manuscript and made numerous valuable comments and suggestions concerning content and layout. Furthermore, we thank Violaine Margueret for proof reading and improving the manuscript, and Julie Ferland for her invaluable help in producing the numerous figures. Finally, we thank the team of Éditions MultiMondes for their excellent editing.

# Foreword

The existence of the symbiotic mycorrhizal association was brought to light in Germany by Frank towards the end of the 19th century, and a little later in France, by Bernard. In Sweden, during the first half of the 20th century, Melin described ectomycorrhizas associated with forest trees, and studied the relationship between these and a wide range of higher fungi (mostly Basidiomycetes). Concurrently, in the U.S.A., Hatch demonstrated the irrefutable role of ectomycorrhizal fungi in the mineral nutrition of forest trees. However, it was not until the middle of the 20th century, that Harley and his research group in England were able to show the physiological mechanisms by which ectomycorrhizal fungi obtain carbon – their energy source – from the hosts they colonize. They also demonstrated the role of ectomycorrhizas in mineral nutrient uptake from the soil. At the same time, Mosse highlighted the role that arbuscular mycorrhizal fungi play in the roots of the majority of plant species, including most agriculturally important crops. By the beginning of the 1970s, it was clear that the formation of mycorrhizas was essential for the survival and growth of most members of the plant kingdom.

Over the last 35 years, numerous scientific studies have clearly highlighted the fundamental role that mycorrhizal fungi play in natural ecosystems, and in those managed by man. However, in spite of the undeniable evidence provided by repeated scientific studies, the majority of horticulturalists, agriculturalists, sylviculturalists and environmentalists, still have only a limited understanding of the importance of mycorrhizas. This major knowledge gap must be bridged before mycorrhizal fungi can fully realize their potential to enhance sustainable plant production.

One of the reasons behind the lack of awareness concerning the importance of the different mycorrhizal symbioses is, without doubt, due to the fact that, in the past, scientists working in this field tended to only talk amongst themselves at national and international conferences. Furthermore, their publications were rarely aimed at publicizing the results of their research to potential users. Although the scientific community has now started to open up to the possible applications of mycorrhizal fungi in different plant production systems, there is still a long way to go.

It is in this optic that we have prepared this book, which aims to help everyone understand the biology of mycorrhizal fungi and the fascinating mycorrhizas that they form with plant roots. This book clearly shows how we can benefit from using these fungi in diverse aspect of plant production, while respecting the delicate balance of nature.

We have sought to address the needs of students and lecturers at all levels, particular those interested in biology, horticulture, agriculture, forestry, and environmental science. This book is also aimed at plant producers, whether they are growing under glass or in the field, and under small- or large-scale production systems. It is also aimed at those conducting reclamation, restoration, and revegetation work, and at garden centre staff. In addition, this book will be a valuable tool for mycologists and harvesters of edible forest mushroom seeking a more in-depth understanding of mycorrhizal fungal biology, and how certain highly sought-after, gastronomically important species can be cultivated.

### THE POST-PHOSPHATE ERA

Just as the global availability of oil has reached a peak, the maximum availability of phosphate reached a peak in the United States of America in 1988, and it has been predicted that the maximum peak in global availability will be reached by the beginning of the 2030s. Because there is no substitute for phosphate, this vital resource will, sooner or later, limit human population growth. Given the role of mycorrhizal fungi in phosphate uptake by plants, their more widespread use in plant production is likely to be an important key to reducing the existing pressures on stocks of this non-renewable resource.

New and exciting discoveries in the field of mycorrhizal research are continually being added to the long list of scientific publications in this domain. We hope that this book will go someway towards condensing and simplifying the existing information, rendering it accessible to the widest possible audience.

# Contents

CHAPTER 1

# Living in symbiosis

The study of animals often brings us swiftly to the conclusion that living organisms are in continual **competition**[1], either directly or indirectly, for food and habitat. Furthermore, many are also subject to more aggressive relationships such as **parasitism** and **predation**.

If we look at higher plants, we also find clear examples of competition, parasitism and predation, and we could stop our reflection there. However, if we go beyond this primary perception, we discover the existence of more harmonious relationships in the plant kingdom and, for that matter, in the animal kingdom, too.

In fact, by putting in common their genomes, ancestral micro-organisms allowed the evolution of **eukaryotes**, which eventually gave rise to the fungi, algae, plants and animals that exist today. Our mitochondria, which are vital for respiration and the release of energy from our food, and the chloroplasts of green plants, which enable them to fix the sun's energy through photosynthesis, both resulted from ancestral **symbioses** between bacteria and early cells (Figure 1.1). It is easy to imagine how chloroplasts developed within ancestral phagocytic protists that fed on algae and bacteria (see **phagocytosis**). Today, there exists a remarkable protozoan, *Cyanophora paradoxa*, which hosts a cyanobacterium in its cytoplasm. The protozoan is remarkable in having two nuclei, one containing its own genetic code and the other, that of the cyanobacterium. When *C. paradoxa* divides, the two nuclei divide synchronously. However, during their long coexistence, the cyanobacterium has lost an essential

1. See glossary for the definition of highlighted terms.

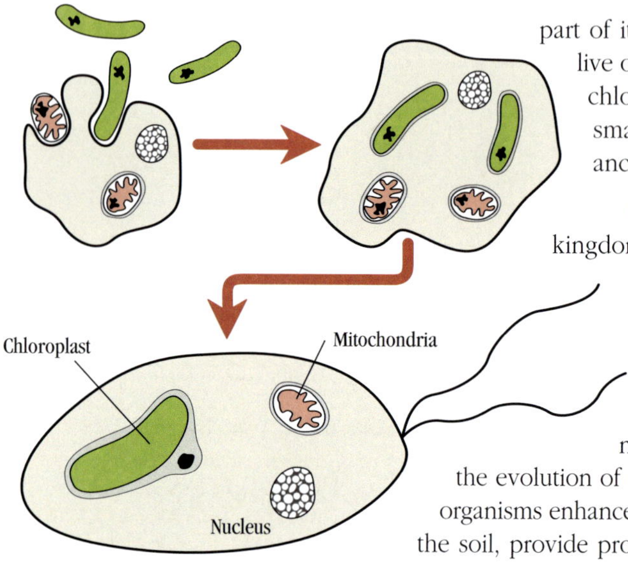


**Figure 1.1** – During the course of evolution, certain cyanobacteria (green) and bacteria (red) that were ingested by ancestral protists managed to survive, and these slowly evolved into the chloroplasts and mitochondria we know today. Each of these organelles conserves a vestige of its origin in the form of chloroplastic or mitochondrial DNA, respectively.

part of its genetic code and is now unable to live outside the protozoan. Interestingly, the chloroplasts of higher plants also carry a small fraction of the genetic code of an ancestral cyanobacterium.

If we examine members of the plant kingdom from a symbiotic angle, we rapidly discover that green plants generally live in close collaboration with numerous soil micro-organisms. We also discover that numerous soil organisms, many invisible to the naked eye, have played a key role in the evolution of the higher plants present today. These organisms enhance plant nutrition and water uptake from the soil, provide protection against pathogenic organisms, and increase the plant's resistance to environmental stressors. These organisms also help shape the composition, distribution and functioning of terrestrial ecosystems.

After having studied this topic closely, one cannot help but arrive at the conclusion that useful micro-organisms are omnipresent, that they interact dynamically with plants at all stages of their development, and that they are essential for the stability of ecosystems. In contrast, organisms that cause damage through parasitism or predation are restrained to a limited number of hosts, and they may even cause the disappearance of the species on which they rely for food. The plant model, with abundant examples of symbioses, clearly shows an often hidden facet of the living world, which differs markedly from the animal model that is generally presented. In this context, it is worth re-examining the different types of interactions between organisms that Robert Whittaker (Figure 1.2) presented.

## DIFFERENT TYPES OF INTERACTION

Whittaker's vision of the different possible interactions between two co-occurring organisms is summarized in the simple diagram shown on the right. In the first type of interaction, shown in the lower right-hand corner, both organisms are in competition for a given resource – for example, two plants in the same pot that need the same limiting mineral nutrients. Under such conditions, both plants are likely to be negatively affected, and exhibit reduced growth and reproductive success. To avoid such situations, evolution has selected for certain strategies based on **antibiosis,** and on **amensalism**. In the former, certain plants or animals produce allelopathic substances or antibiotics, which poison or kill neighbouring plants. In the case of amensalism, a dominant organism deprives its neighbour of a needed resource – for example, a tree shading out smaller plants in the understorey. In what are considered to be more evolved interactions, the relationship may be more aggressive, and can take the form of parasitism or predation. In the case of parasitism, one of the two organisms is weakened or killed, and occasionally the host may be virtually eradicated. A good example of this is Dutch elm disease, which has recently decimated the American elm population. In North America, beetles accidentally introduced from Europe, transfer the pathogenic fungus to uninfected host trees. Successful parasites are often highly host specific; however, when taken to the extreme, these organisms can eliminate the host on which their own survival relies. Other organisms succeed in co-occurring without any form of aggression towards their associates. For example, Cattle Egrets associate with large ruminants, feeding on the insects they disturb, but apparently give nothing in return. This type of relationship is known as **commensalism**. The study of interspecific interactions suggests that in plants, the evolutionary tendency is to move from competition to parasitism, passing by antibiosis, and from parasitism to symbiosis passing via commensalism. In the animal kingdom, the evolutionary tendency is to move towards predation, passing via amensalism, and then symbiosis passing via commensalism.

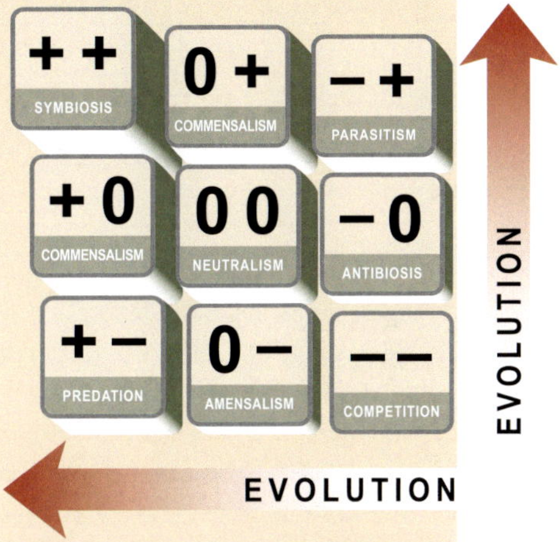


**Figure 1.2** – Co-occurring organisms may interact in different ways, with competition being considered the most primitive form, and symbiotic relationships, the most evolved. In the plant kingdom the evolutionary intermediate between these extremes is often parasitism, while in the animal kingdom it tends to be predation.

> Globally, Whittaker's work suggests that predation and parasitism constitute more primitive means of escaping competition, whereas mutualistic symbioses appear to be a more evolved and successful evolutionary process.
>
> According to Darwin's theory of evolution, natural selection favours those organisms that join forces.

In light of evidence from the plant kingdom, a new look at the animal kingdom, which has typically been associated with relationships based on competition, parasitism and predation, will no doubt lead to the discovery of a surprising number of mutualistic associations. These will add to well known examples, such as the *Hydra*, a relative of the jellyfish, which owes its green colouration to a symbiotic alga that is contained within its tissue. The same is true for corals, which each host a different algal species. Furthermore, termites and ruminants, two of the planet's important recyclers, are both able to obtain nutrients from cellulose because they contain a complex symbiotic gut flora comprising protozoa in symbiotic association with bacteria.

Scientific research conducted over the last 50 years has provided a better understanding of the complex symbiotic relationships between plants and their soil-borne microbial associates. This notion is essential for the understanding and sustainable use of the natural environment, and is essential for ensuring the conservation of its different components. Furthermore, by considering plant systems from a symbiotic perspective, we are better able to understand factors driving plant survival and productivity. In turn, this has allowed the development of new concepts and technologies in cropping systems that are more respectful of the environment. The understanding and use of these symbioses is the primary key to sustainable agriculture.

In the next chapter, we will briefly outline some of the principal symbioses found in the plant kingdom, and give a brief description of their microscopic structure, morphology, physiology and ecological roles.

CHAPTER 2

# Plant symbioses and their structures

Table 2.1 summarizes the form and function of the main types of symbiotic associations found in the plant kingdom, and outlines their relative importance in the natural environment.

**TABLE 2.1**

**THE PRINCIPAL SYMBIOSES WITHIN THE PLANT KINGDOM**

| SYMBIOSIS | NATURE OF THE MICROBIAL SYMBIONT | PLANTS INVOLVED | MICROBIAL STRUCTURE | PERCENTAGE OF PLANT SPECIES | HOST PLANT STRUCTURE | FUNCTIONS ACQUIRED OR ENHANCED |
|---|---|---|---|---|---|---|
| **Lichens** | Ascomycetes and Basidiomycetes | Green algae or cyanobacteria | Mycelium surrounding algal cells or cyanobacteria | n/a | Algal cells or cyanobacteria surrounded by fungal mycelium | Mineral nutrition, water supply, N fixation, resistance to desiccation |
| **Bacteriorhiza** | Bacteria in the genera *Rhizobium* and *Bradyrhizobium* | Legumes such as beans, lucerne and Acacia | Bacteroids in the cortical cells of the roots | 5% | Root nodules, often short lived, producing leghemoglobin | Fixation of atmospheric nitrogen |
| **Actinorhiza** | Actinomycetes (filamentous prokaryotes) in the genus *Frankia* | Diverse genera including *Alnus*, *Myrica*, *Dryas*, *Casuarina* | Mycelium, septate vesicles in cells of the root cortex | 1% | Perennial nodules lacking leghemoglobin | Fixation of atmospheric nitrogen |
| **Phycorhiza** | Cyanobacteria | Members of the Cycadales, such as the genus *Cycas* | Intracellular cyanobacteria in cells of the root cortex | <1% | Dichotomous branching of roots, negative geotropism | Fixation of atmospheric nitrogen |
| **Mycorrhizas** | Ascomycetes, Basidiomycetes, Glomeromycetes | Numerous vascular plants | Mycelium associated with the root | >85% | Root-fungal complex | See Table 2.2 |

## Lichens

The lichens are complex organisms comprising a microscopic alga and a filamentous fungus growing in symbiosis. The fungi are typically Ascomycetes, but a small percentage of lichens involve Basidiomycetes. The life cycles of these two fungal groups are shown in Figures 2.1 and 2.2, respectively.

Land plants face a number of common challenges. As photoautotrophs, plants must be able to capturing the sun's energy and transform it, via the process of photosynthesis, into sugars, which serve as a chemical energy source. They must also be able to obtain water and sufficient mineral nutrients, such as nitrogen, phosphorus and potassium, for growth and function. Photosynthesis only occurs in chloroplast-containing organisms; however, a group of chemoautotrophic bacteria produce organic compounds using chemical energy obtained from oxidizing certain inorganic substances.

In the case of lichens, the photosynthetic partner is a unicellular or filamentous alga, belonging to one of a limited number of species. In an aquatic environment, algal cells are bathed in a weak nutrient solution that generally contains all the minerals needed for growth. In contrast, in terrestrial environments, water and minerals are contained within the soil matrix, where light cannot penetrate and where photosynthetic algae cannot survive. The logical solution to this problem is for the alga to form an association with a non-photosynthetic organism capable of growing in the dark, and able to efficiently explore the soil, and other substrates, for water and mineral nutrients. Multicellular fungi, with their long thread-like filaments or hyphae, meet this requirement, making them a natural ally for algae. The association between the two partners allows the algae to survive in exposed terrestrial environments.

Lichens are highly successful and able to grow on that most challenging of substrates, exposed rock, where they generally grow as crusts (crustose). However, lichens are also found in forests, and in these more protected sites they may proliferate as thin leaf-like sheets (foliose), or miniature tree-like structures (fruticose) (Figure 2.3). The microscopic examination of a lichen thallus reveals the presence of an alga, most often unicellular, surrounded by a matrix of fungal filaments (Figure 2.4). Observation at a higher magnification, would allow us to follow the fungal hyphae into the adjacent substrate, were they absorb water and mineral nutrients from soil pores and cracks in the rock.

## FIGURE 2.1 – SCHEMATIC DRAWING OF THE LIFE CYCLE OF AN ASCOMYCETE

Peziza

Morel

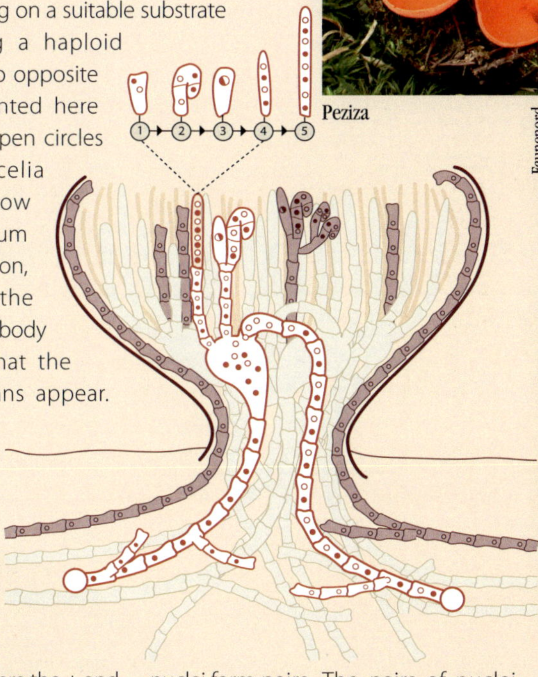


Members of the Ascomycetes, such as species in the genus *Peziza*, often produce small cup- or saucer-shaped spore-producing structures or fruitbodies. However, other members of this class, such as the Morels (genus *Morchella*), form larger upright fruitbodies. In the Ascomycetes, sexual reproduction occurs in asci. These generally elongated structures produce haploid spores, which act as both propagules and gametes. After dispersion, spores landing on a suitable substrate germinate, producing a haploid mycelium of one of two opposite mating types, represented here by full circles (+) and open circles (–). Compatible mycelia of opposite types grow together in a primordium without undergoing fusion, and it is only once the development of the fruitbody has been triggered, that the male and female organs appear. In the ascogonium, the female nuclei (+) multiply by mitosis and form a coenocyte. The male mycelium (–) develops a multi-nucleic antheridium, and injects its nuclei into the ascogonium, where the + and – nuclei form pairs. The pairs of nuclei become isolated in bi-nucleic compartments (dikaryons) ① until the formation of the mother cell from which the ascus develops. The cell divides to form the crozier ② and the nuclei fuse together to form a zygote, the only truly diploid cell in the entire cycle ③. The zygote immediately undergoes meiosis to produce four haploid nuclei (two + and two –) ④. A further division of each nucleus gives rise to the formation of eight ascospores (four + and four –) ⑤.

## FIGURE 2.2 – SCHEMATIC DRAWING OF THE LIFE CYCLE OF A BASIDIOMYCETE

Henri-Paul Giguère

Red-belted Polypore

Members of the Basidiomycetes vary greatly in shape and size. The Horse Mushroom and the Red-belted Polypore are two examples. The reproductive structures of fungi in this class are referred to as basidia; in the case of the Horse Mushroom, these line the gills formed under the cap of the fruitbody, and in the case of the Red-belted Polypore, they line the inside of tubes on the underside of the fruitbody. Each basidium carries four haploid spores, 2 + and 2 –, which act as both propagules and gametes. After dispersion, spores landing on a suitable substrate germinate, producing a haploid mycelium of one of two opposite mating types, represented here by full circles (+) and open circles (–). Soon after germination, compatible mycelia of opposite types fuse to give a dikaryotic mycelium of long filaments (hyphae) comprised of cells containing two compatible but unfused nuclei. In contrast to the Ascomycetes, the whole of the fruitbody is made up of such hyphae. Within the gills or pores, the nuclei in the dikaryotic cell situated at the tip of each hypha ① fuse to form a young diploid basidium ②. Meiosis in the developing basidium results in the production of four nuclei, 2 + and 2 –. The nuclei then migrate out of the basidium to form four basidiospores ③, which are then ejected from the basidium, and the life cycle starts over again.

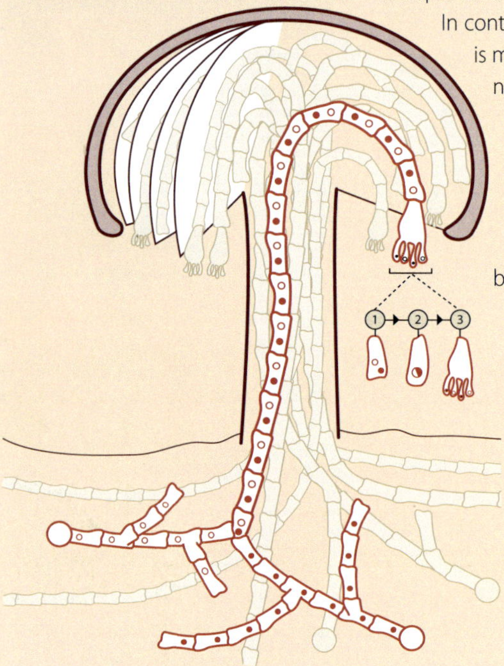

Horse Mushroom

Jacques Landry

**Figure 2.3** – Although the primitive and coastal lichens are typically crustose (a), in other environments foliose (b) and fruticose types (c) are frequently encountered.

Photos: a: J. André Fortin, Université Laval; b and c: Serge Payette, Université Laval

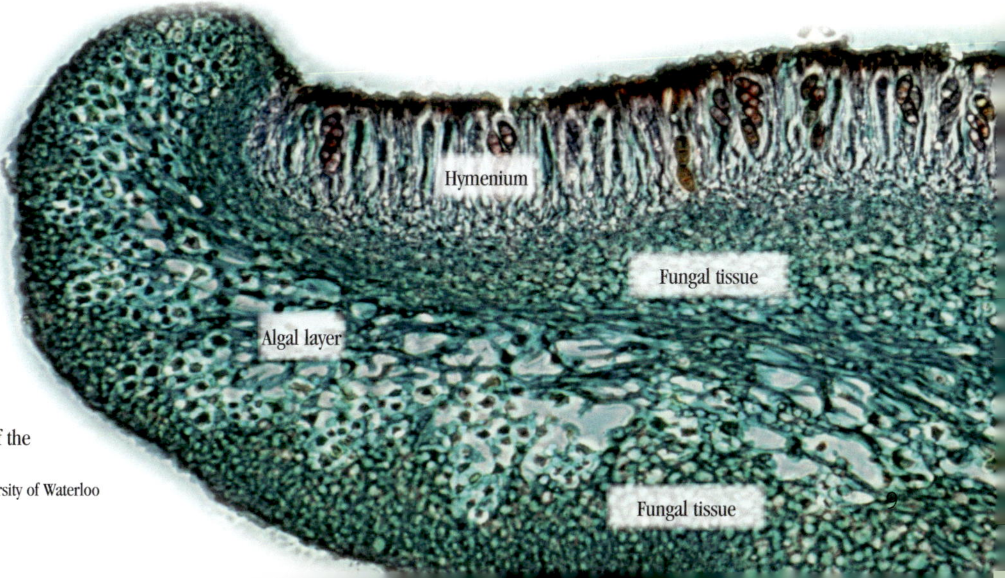


**Figure 2.4** – A vertical section through a lichen showing the hymenium; note the algal cells and the fungal tissue with asci, which are typical of the Ascomycetes.

Photo: Bryce Kendrick, University of Waterloo

The lichens are a perfect example of a plant symbiosis, where the roles of photosynthesis, and of water and mineral nutrient uptake, are shared in a complementary manner between two distinctly different organisms. The alga, by means of its photosynthetic capacity, provides the fungus with an energy source in the form of sugars, while the fungus ensures a supply of water and mineral nutrients for the two associated organisms. Although only a few dozen species of algae form lichen, several thousand species of lichenized fungi have been described and most are Ascomycetes. The evolutionary origin of lichens, their ecological role and their utility will be examined in the next chapter.

## Mycorrhizas

Mycorrhizas, like lichens, are symbiotic associations between a photosynthetic organism, this time a green plant, and a filamentous fungus. The term mycorrhiza is a mix of both Latin and Greek, and literally means fungus-root (*μukeς* = fungus, *rhiza* = root). There are various mycorrhizal associations (Table 2.2), but the four most common ones are arbuscular mycorrhizas, ectomycorrhizas, ericoid mycorrhizas, and orchid mycorrhizas. Almost all terrestrial green plants form mycorrhizas, and only members of a few families, such as the Chenopodiaceae and the Cruciferae, are typically non-mycorrhizal.

The mycorrhizal symbiosis is a fundamental and universal phenomenon among the vascular plants and bryophytes, which has existed since the first plants appeared on dry land more than 450 million years ago. As in the case of the algae and fungi that form lichens, the photosynthetic partner in the mycorrhizal symbiosis produces sugars for both organisms, and the fungus absorbs enough water and mineral nutrients from the substrate to meet its needs and those of its host.

The fungi that form the different types of mycorrhizas belong to diverse classes, and this aspect will be investigated in greater detail as we discuss each mycorrhizal type. As the roots of most plants are structurally similar, before we go any further it will be useful to quickly look at the structure and organization of a typical plant root (Figure 2.5). The root elongates by means of its apical meristem, thus allowing it to explore the soil for water and mineral nutrients. The root hairs play an active role in water absorption and the principal zone for uptake is situated between the meristem and the start of the suberized zone. The latter is generally characterized by a much reduced water and mineral nutrients absorption capacity. Therefore the total surface area available for absorption is relatively small with regards to that of

**TABLE 2.2**

| THE DIFFERENT MYCORRHIZAL ASSOCIATIONS | | | | | |
|---|---|---|---|---|---|
| **MYCORRHIZAL TYPE** | **FUNGI INVOLVED** | **HOST PLANT** | **FUNGAL STRUCTURES** | **HOST STRUCTURES** | **PHYSIOLOGICAL IMPACT** |
| **Arbuscular** | Microscopic fungi: Glomeromycetes: ~ 200 species | Bryophytes and vascular plants: 70% of existing species | Arbuscules and intracellular vesicles, extraradical mycelium and spores | Little change, yellow colouration | Increased access to P and Zn, resistance to drought, disease and grazing, marked changes in metabolism |
| **Ectomycorrhiza** | Higher fungi: Ascomycetes and Basidiomycetes: several thousand species | Gymnosperm and angiosperm trees: 5% of existing species | Fungal sheath, intercellular mycelium, rhizomorphs, sclerotia, ascomata, basidiomata. No intracellular penetration | Cortical hypertrophy, dichotomous branching or pinnate | Increased access to mineral nutrient, use of organic nitrogen, resistance to disease and nematodes, tolerance to low pH and heavy metals |
| **Ectendomycorrhiza** | Deuteromycetes: a few species | Pines: rare | Thin sheath, intercellular mycelium, intracellular penetration | Cortical hypertrophy, ramifications | idem |
| **Arbutoid** | Basidiomycetes: a few species | Ericaceae: rare | Thin sheath, intracellular penetration, basidiomata | Cortical hypertrophy | idem |
| **Ericoid** | Ascomycetes: several tens of species | Ericaceae: 5% of existing species | Intracellular mycelium, ascomata | Little modification | idem |
| **Orchid** | Basidiomycetes and little studies sterile mycelia | Orchidaceae: 10% of existing species | Intracellular mycelium in dense pelotons, basidiomata | Little modification | Often essential for morphogenesis, saprophytic nutrition of the plant, protection against pathogens |
| **Sebacinoid** | *Piriformospora*; Basidiomycetes: a few species | Various | Intracellular mycelium | Little modification | Little known |

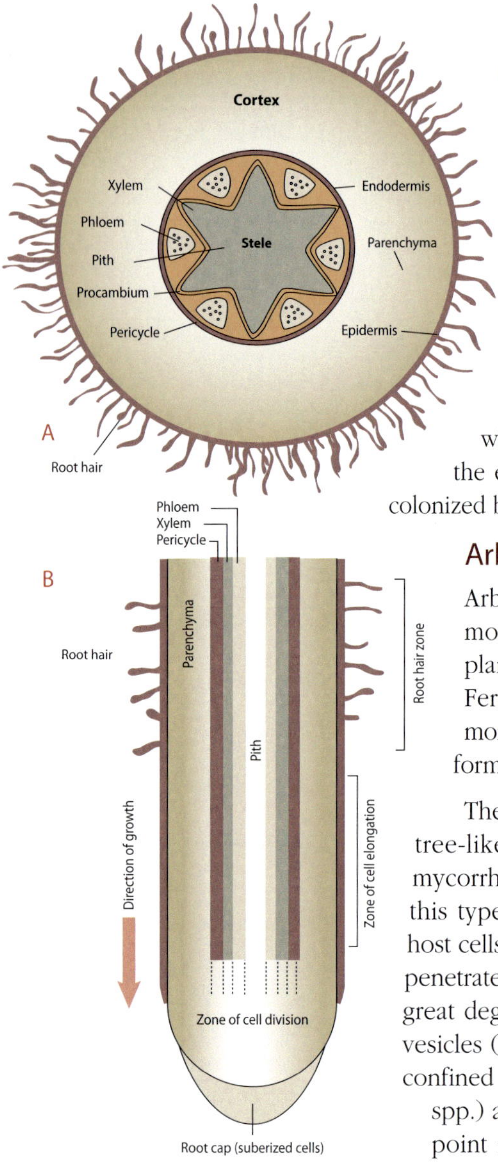


**Figure 2.5** – Transverse (a) and longitudinal (b) section of a primary root.

the entire root system, and of the plant as a whole. Typically, in the centre of the root there is a vascular cylinder comprising ascending vessels (xylem) that conduct water and mineral nutrients from the roots to the aerial parts of the plant, and descending vessels (phloem) that carry a part of the sugars produced by photosynthesis towards the roots. The central cylinder is surrounded by the endoderm, which is a layer of cells that are able to preferentially select the mineral nutrients useful for plant growth and, to a certain degree, exclude others which are not or which may be harmful. Between the outer epidermis and the endoderm is the cortical parenchyma, which may be colonized by the mycelium or hyphae of mycorrhizal fungi.

## Arbuscular mycorrhizas

Arbuscular mycorrhizas (Figure 2.6) are formed by more than 70% of the existing species of vascular plants, as well as by numerous mosses and liverworts. Ferns, lycopods, several conifers and the majority of monocotyledonous and dicotyledonous flowering plants form this type of mycorrhizal association.

The term "arbuscular" refers to a unique microscopic tree-like structure (Figure 2.7 a and b) that arbuscular mycorrhizal fungi develop in the cortical cells of roots. In this type of mycorrhiza the fungus does not surround the host cells, as it does in the case of ectomycorrhizas, instead it penetrates them, but without disturbing their structure to any great degree. Furthermore, certain species form intercellular vesicles (Figure 2.7c). The development of these fungi is not confined to the root, as in the bacteriorhizas (e.g., *Rhizobium* spp.) and the actinorhizas (*Frankia* spp.). From an anchor point in the root, arbuscular mycorrhizal fungi develop an extraradical phase in the soil. This mycelium forms a

FIGURE 2.6 – SCHEMATIC DRAWING OF A ROOT APEX COLONISED BY AN ARBUSCULAR MYCORRHIZAL FUNGUS

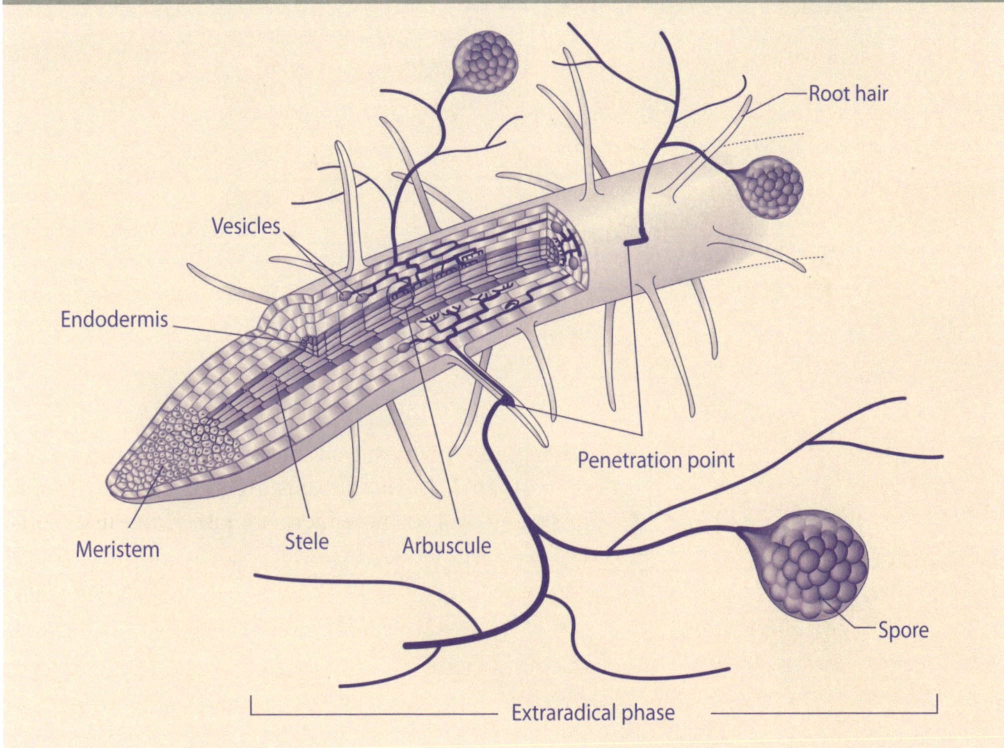


network that penetrates the adjacent soil in all directions. This very fine mycelium (Figure 2.7d) provides a considerable surface area that is in contact with the soil. It has been estimated that the surface area of the mycelium of arbuscular mycorrhizal fungi in 1 $m^2$ of prairie soil, is roughly 90 $m^2$, and one kilometre of hyphae may be associated with a single leek plant growing in a one-litre pot. This mycelium can access water and mineral nutrients in the smallest of cracks and pores in the substrate.

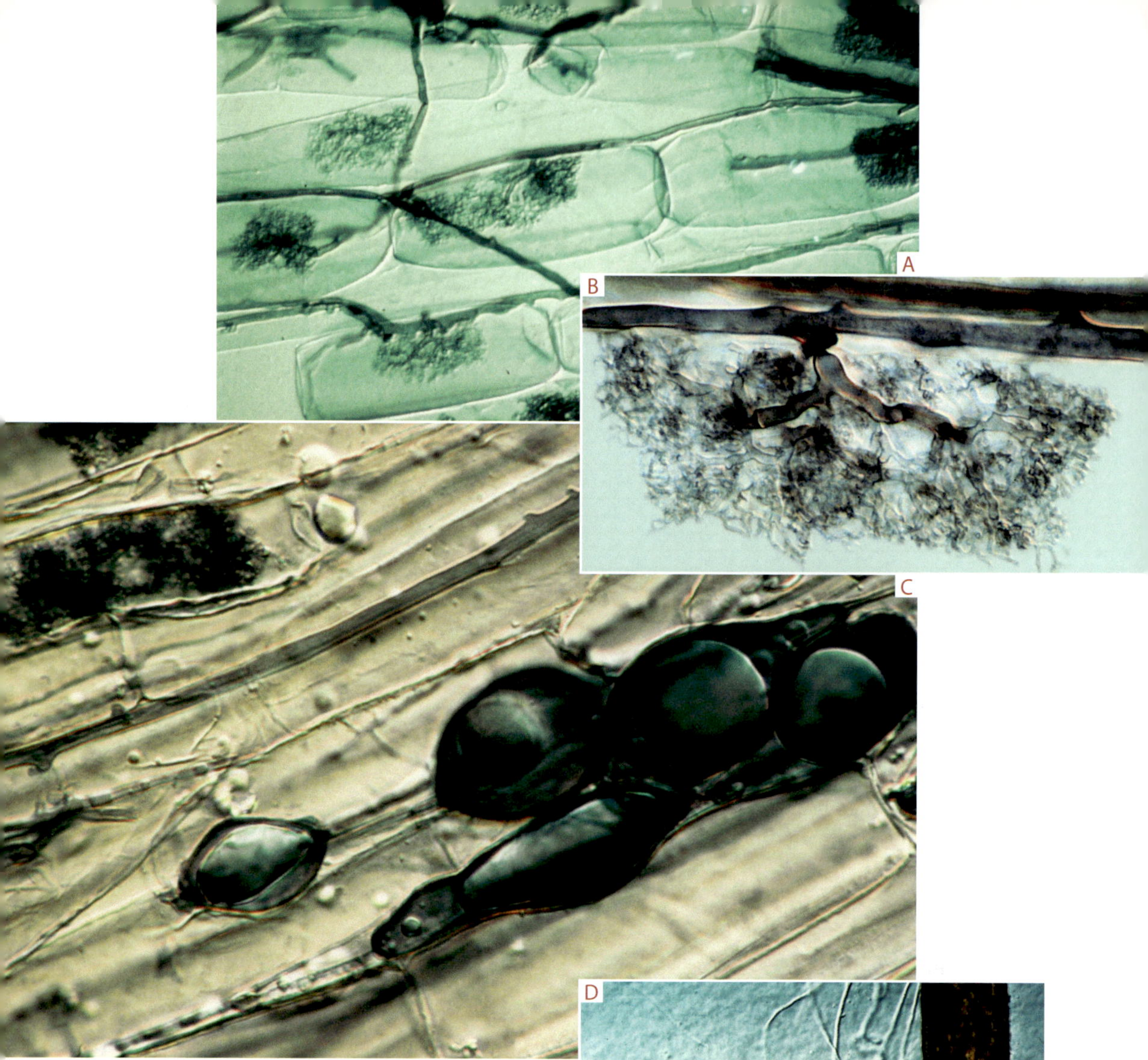

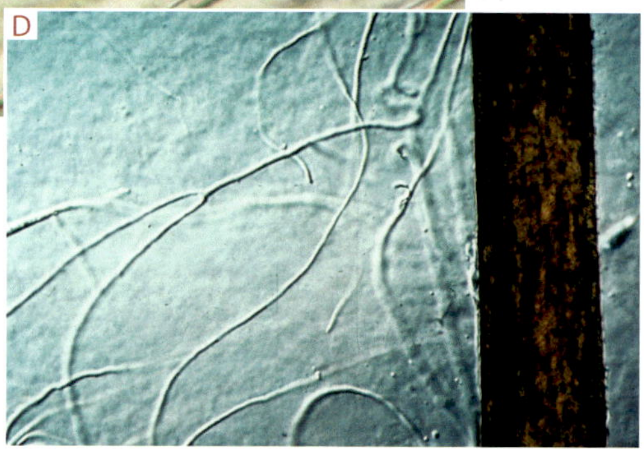

**Figure 2.7** – Characteristic structures of arbuscular mycorrhizas. (a) Intercellular mycelium and intracellular arbuscules. (b) Detailed structure of an arbuscule. (c) Intercellular vesicles. (d) The diameter of extraradical hyphae, compared to that of a much larger human hair (100 µm).

Photos: a and c Larry Peterson et al., CNRC; b: Mark Brundrett et al., *Canadian Journal of Botany*, CNRC; d: George Barron, University of Guelph

The mycelium of arbuscular mycorrhizal fungi is coenocytic (see **coenocyte**), that is, it is not divided into numerous individual cells by cross walls or septa. Therefore, the nuclei are not singly confined in individual well-defined structures, but occur together in the cytoplasm of the long cylindrical hyphae. This gives rise to a network that resembles a huge three-dimensional spider's web. The importance of this arrangement of hyphae becomes obvious when the symbiotic functioning of arbuscular mycorrhizas is considered. Microscopic investigation of the network reveals that the tubular hyphae are interconnected and that there is free movement of cytoplasm between the different interconnected hyphal elements. Closer observation reveals that this cytoplasmic streaming or **cyclosis** is bidirectional.

Arbuscular mycorrhizal fungi produce spores (Figure 2.8) at intervals along the length of the coenocytic hyphae growing in the soil, and in some species the spores are grouped into simple sporocarps. Spores provide a reserve of fungal propagules within the soil that can also serve to disseminate the species. Certain species also produce intraradical reproductive structures known as vesicles (Figure 2.7c). These are formed in the root cortex, and have similar properties to spores. Indeed, spores, vesicles and root fragments containing vesicles, can all serve to propagate a given species and it is for this reason that they are often simply referred to as propagules.

Compared to the spores of other fungi, which are generally between 5 and 20 micrometers (μm) in diameter, those of arbuscular mycorrhizal fungi reach impressive sizes and commonly range between 40 and 500 μm. In comparison, the diameter of a human hair is approximately 100 μm (Figure 2.7d). These spores also possess extremely thick walls (2-35 μm) that protect them against environmental stresses, and increase their longevity. It has been shown that arbuscular mycorrhizal fungal spores ingested by small rodents pass unharmed through their digestive tract, which favours their dispersion under natural conditions.

Recent morphological and phylogenetic studies have resulted in all the species of arbuscular mycorrhizal fungi being grouped into a new branch, the Glomeromycota, which presently contains four orders, eight families, 12 genera and approximately 200 species. This relatively small number of species forms mycorrhizal associations with approximately 225,000 species of land plants.

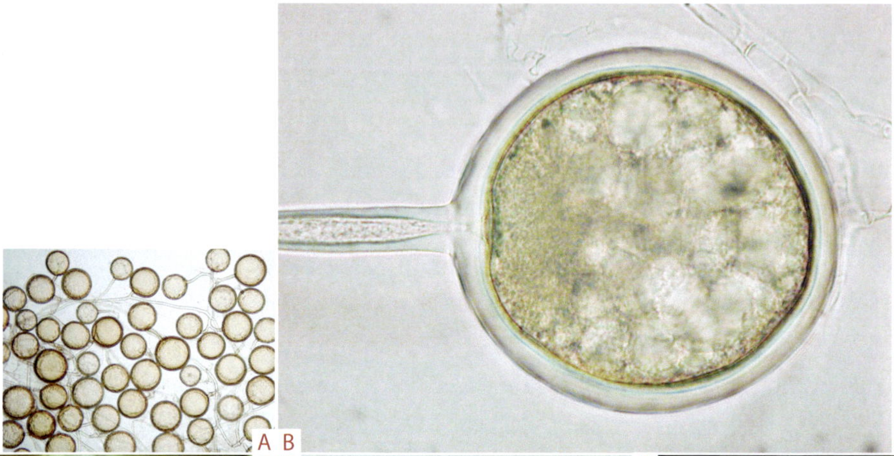


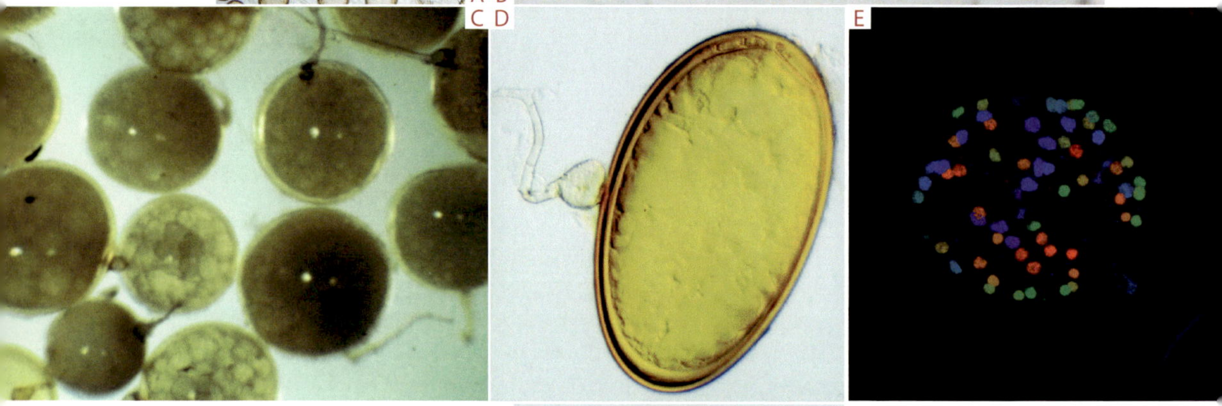


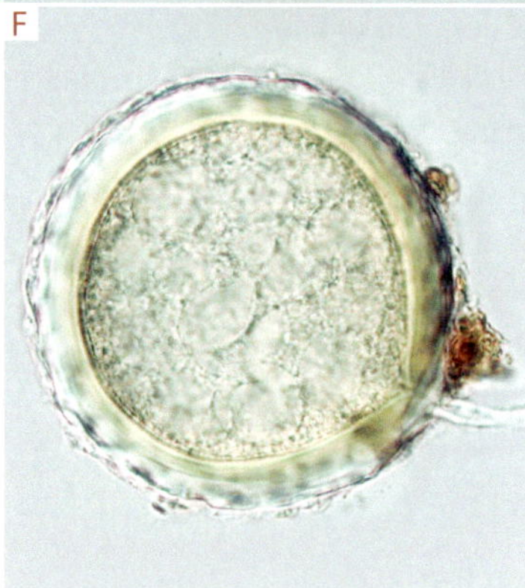


**Figure 2.8** – Spores of a range of arbuscular mycorrhizal fungi. (a) Spores of *Glomus intraradices*. (b) Spore of *Glomus clarum* showing a straight suspensor hypha. (c) Spores of *Scutellospora calospora* with lipid droplets clearly visible in the cytoplasm. (d) Spore of *Scutellospora calospora* showing a bulbous suspensor hypha. (e) Spore of *Glomus diaphanum* containing several genetically distinct nuclei. (f) Spore of *Glomus lamellosum* showing a very thick spore wall.

Photos: a, b, d and f: Yolande Dalpé, Agriculture and Agri-Food Canada; c: Valentin Furlan, Université Laval; e: Mohamed Hijri, Montreal University

Arbuscular mycorrhizal fungi are obligate symbionts. At present, it is impossible to grow them in pure culture in the absence of a host plant root. In the laboratory, and more recently on an industrial scale, this problem has been partially overcome by growing these fungi in Petri dishes in association with host plant root-organ cultures. This subject will be considered in greater depth in Chapter 8, which covers arbuscular mycorrhizal fungal **inoculum** production.

Figure 2.9 compares the structures of a typical arbuscular mycorrhiza with those of a typical ectomycorrhiza. This latter group will be covered in the next section.

## FIGURE 2.9 – SCHEMATIC COMPARISON OF AN ARBUSCULAR MYCORRHIZA AND AN ECTOMYCORRHIZA

In arbuscular mycorrhizas, the mycelium of the fungus penetrates the cortical roots cells; however, in ectomycorrhizas, the mycelium surrounds, but does not penetrate, these cells.

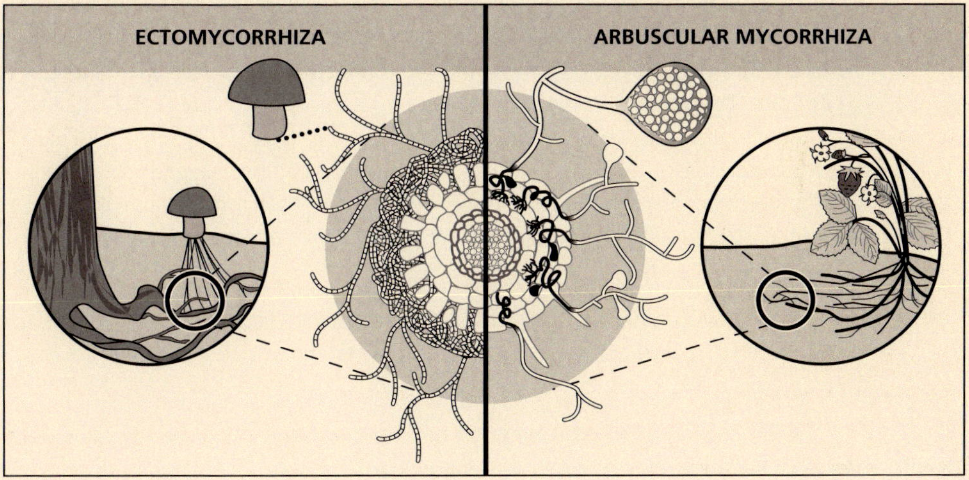

## The ectomycorrhiza

In evolutionary terms, ectomycorrhizas are much more recent (~250 million years ago) than arbuscular mycorrhizas, and ectomycorrhizal fungi form symbiotic associations with a much smaller group of land plants. Approximately 5% of vascular plant species (25,000) live in symbiosis with ectomycorrhizal fungi, which is in marked contrast to the 70% of vascular plants that form arbuscular mycorrhizal associations. Nevertheless, ectomycorrhizal plants occupy huge geographical areas, including the circumpolar boreal forest. Although a number of ectomycorrhizal fungi belong to the Ascomycetes, the majority of the mycobionts belong to the Basidiomycetes (Figures 2.1 and 2.2). Many of the higher fungi that form ectomycorrhizal associations are well known to mycologists because their fruitbodies only grow on the forest floor in the vicinity of their host trees. Depending on the tree species colonized, the form of the ectomycorrhizal root can be simple (Figure 2.10d), pyramidal (Figure 2.10a) or dichotomously branched (Figures 2.10b and c).

In contrast to arbuscular mycorrhizal fungi, the hyphae of ectomycorrhizal fungi do not penetrate the cells of their host, but rather develop around the outside of the cortical cells (Figure 2.10e). The tightly packed hyphae form a sheath around the short roots and also penetrate into the root, but are confined to the intercellular spaces of the root cortex. The network formed here (Figure 2.10f) is known as the Hartig net, after the scientist who first described it. From this anchor point in the root, the mycelium explores the adjacent soil forming mycelial cords, comprised of linear aggregates of hyphae, and/or sclerotia, which develop from tightly entwined ball of mycelium. Together, the mycelium of a fungus constitutes a colony and this may occupy several square metres of forest soil.

Unlike the Glomeromycetes that form arbuscular mycorrhizas, the hyphae of Ascomycetes and Basidiomycetes are not coenocytic. Instead, their hyphae are divided by numerous septa.

Under the correct environmental conditions, these dikaryotic hyphae (made up of bi-nucleic cells) give rise to **primordia**, which may subsequently develop into mature fruitbodies. In a developing ascus (Ascomycetes) or basidium (Basidiomycetes) (Figure 2.1 and 2.2) within the fruitbody, the two nuclei of the dikaryon fuse (caryogamy) and immediately undergo meiosis, resulting in spores with a single nucleus (haploid). Under suitable conditions, the spores germinate and produce a primary mycelium, which starts the lifecycle over again.

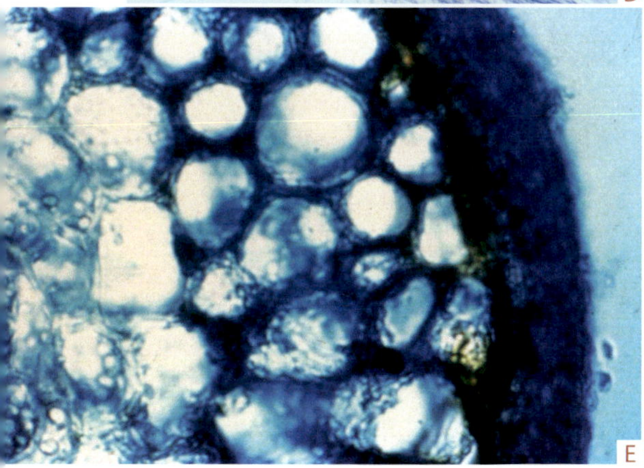

**Figure 2.10** – (a) An exceptionally highly branched pyramidal ectomycorrhiza formed on a balsam fir root. (b) Dichotomic ectomycorrhizas formed on a pine root. (c) Ectomycorrhizas formed by *Thelephora terrestris* (closely related to the polypores) on a pine root. (d) Simple ectomycorrhizas formed by *Cenococcum geophilum*, an Ascomycete. (e) Transverse section of a balsam fir ectomycorrhiza. The fungus has been stained blue to show the external fungal sheath and the Hartig net formed between the cortical cells. (f) Anatomical details of an ectomycorrhiza showing the fungal sheath, the Hartig net, the cortical cells and the endodermis.

Photos: a, e and f: J. André Fortin, Université Laval; b: Larry Peterson et al., CNRC; c: Bratislav Zak, USDA; d: Linda Tackaberry, University of Guelph

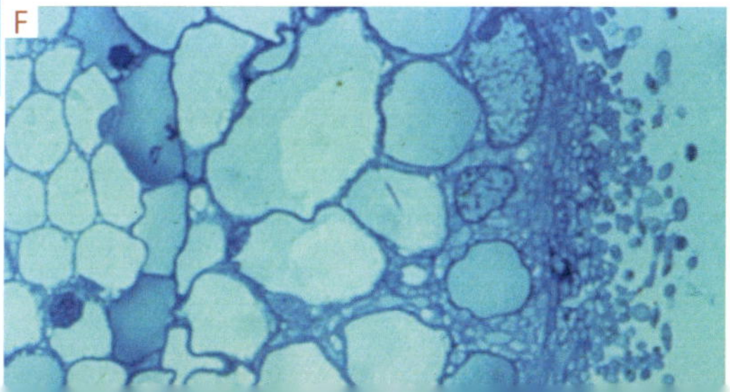

While only a couple of hundred species of fungi form arbuscular mycorrhizas, several thousand species form ectomycorrhizas. These belong to a wide range of genera and include Amanitas, most Boletes, some Tricholomas, Milk-caps, Russulas, Chanterelles Horns of Plenty, and Truffles.

Like the arbuscular mycorrhizal fungi some ectomycorrhizal fungi, such as the Russulas, cannot be grown in **axenic culture**. In contrast, many Boletes, Hebelomas, and Laccarias grow well in pure culture under laboratory conditions; however, the Amanitas, Chanterelles, Milk-caps, and Truffles tend to be more difficult.

## The ericoid mycorrhiza

Fewer vascular plants form ericoid mycorrhizas than form arbuscular mycorrhizas or ectomycorrhizas. Of the few thousand plants that do form this type of mycorrhizal association, most grow in peat-rich acidic environments. These plants all belong to the Ericales, and mainly to the family Ericaceae, notably blueberries and cranberries (*Vaccinium* spp.) and rhododendrons (*Rhododendron* spp.). In addition, some of the other members of the Ericales form **arbutoid** mycorrhizas and other ectomycorrhizas (Figure 2.11).

The fine roots of the Ericaceae generally only have a single layer of cells that constitutes both the epidermis and the cortex. The fungi forming ericoid mycorrhizas belong to the Ascomycetes and certain form small disc-shaped fruitbodies. The fungi penetrate the epidermis of the host root and forms dense hyphal coils (pelotons) in the cells. The mycelium also radiates into the adjacent soil. These fungi are easy to culture under axenic conditions.

Certain species of fungi occurring in the roots of members of the Ericaceae are potentially pathogenic for ectomycorrhizal trees. These fungi are characterised by a heavily pigmented septate mycelium. These colonize tree roots and form intracellular structures that appear to have no advantage for the tree. These little-studied associations are known as pseudomycorrhizas.

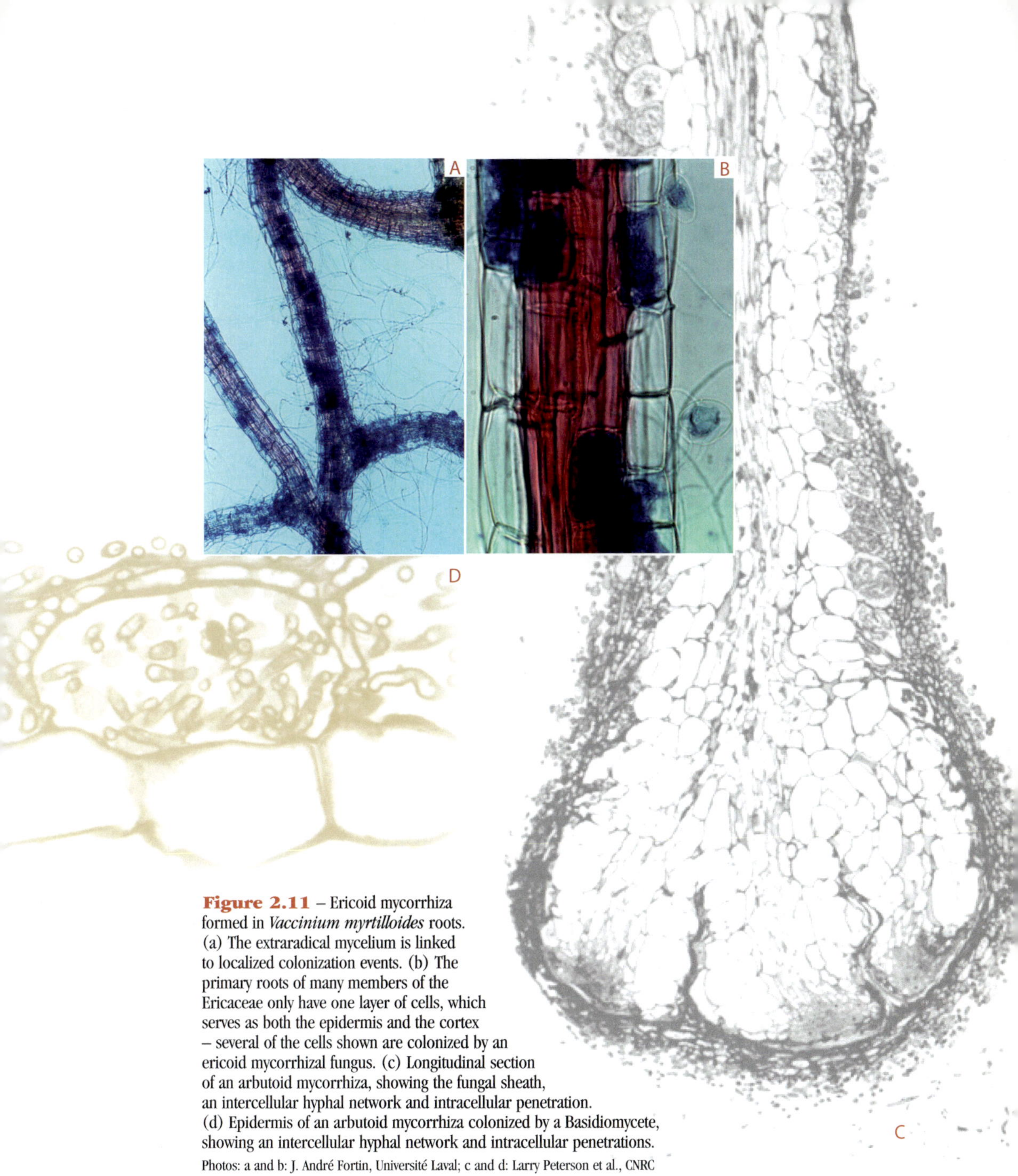


**Figure 2.11** – Ericoid mycorrhiza formed in *Vaccinium myrtilloides* roots. (a) The extraradical mycelium is linked to localized colonization events. (b) The primary roots of many members of the Ericaceae only have one layer of cells, which serves as both the epidermis and the cortex – several of the cells shown are colonized by an ericoid mycorrhizal fungus. (c) Longitudinal section of an arbutoid mycorrhiza, showing the fungal sheath, an intercellular hyphal network and intracellular penetration. (d) Epidermis of an arbutoid mycorrhiza colonized by a Basidiomycete, showing an intercellular hyphal network and intracellular penetrations.

Photos: a and b: J. André Fortin, Université Laval; c and d: Larry Peterson et al., CNRC

Before considering orchid mycorrhizas, Figure 2.12 provides a schematic comparison of the structures of ericoid and orchid mycorrhizas.

In ericoid mycorrhizas, which are typically formed by Ascomycetes, the pelotons of mycelium are confined to the layer of cells that makes up the epidermis and cortex. In contrast, in orchid mycorrhizas, which are normally formed by Basidiomycetes, the fungus colonizes cells across the whole of the cortex, including the epidermis. However, the fungi involved never cross the endodermis.

**FIGURE 2.12 – SCHEMATIC COMPARISON OF ERICOID AND ORCHID MYCORRHIZAS**

In the ericoid mycorrhiza, the roots have only one or two layers of cortical cells that the fungus can colonize; however, in the case of the orchid mycorrhiza, the cortical zone is comprised of several layers of cells and the fungus forms a network of mycelium and pelotons that is much more complex.

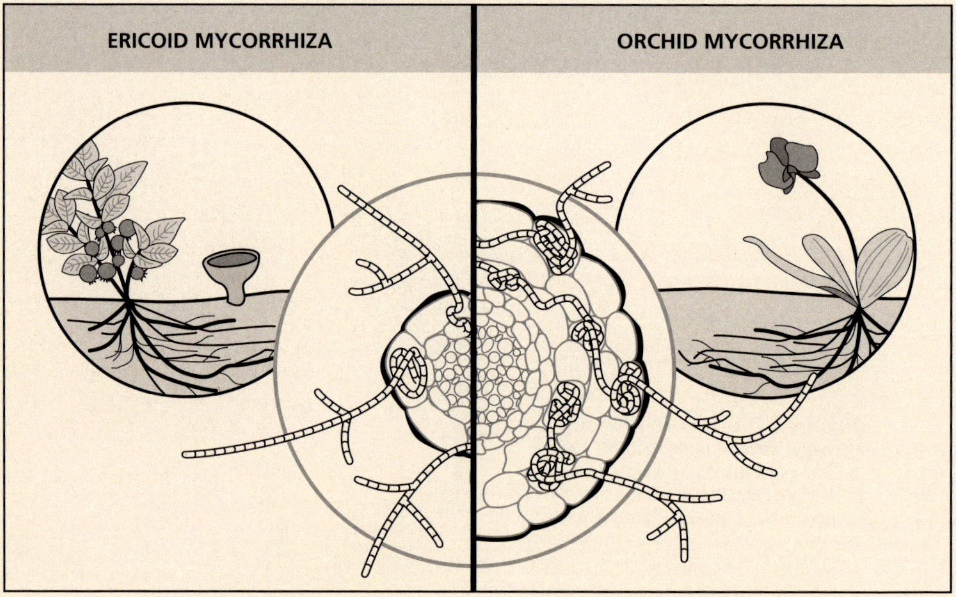

## Orchid mycorrhizas

The use of the term 'mycorrhiza' to describe the symbiosis between orchids and fungi was already widespread by the beginning of the 20th century. Indeed, Noël Bernard discovered that the only way to ensure the successful germination of the minute seeds of orchids under glass was to use soil taken from the vicinity of the mother plant. This soil was found to contain propagules of fungi that were essential for the germination and development of the plant. We now know that this association is rather more complex. In nature, the successful development of a young orchid plant from seed is directly linked to colonization by certain mycorrhizal fungi; however, in the adult phase, other symbiotic fungi take over the role of nutrient uptake. Occasionally, orchid mycorrhizas are essential for growth of the young plant during a certain period, but the symbiotic association is lost once the plant reaches maturity. Certain achlorophyllus orchids depend entirely on their associated fungi for food (Figure 2.13). For example, the achlorophyllus orchid *Gastrodia elata* grows deep in the soil, where it is colonized by the mycelium of a species of *Armillaria*. The fungus in question is a saprophyte that is able to decomposing woody debris. The orchid obtains its carbon source by phagocytically assimilating some of the fungal cells (Figure 2.14). While this is considered a symbiotic association, it may turn out to be a clear case of predation, similar to that observed in some carnivorous plants.

**Figure 2.13** – The coral root is an achlorophyllus myco-heterotrophic orchid.
Photo: Maurice Lalonde, Université Laval

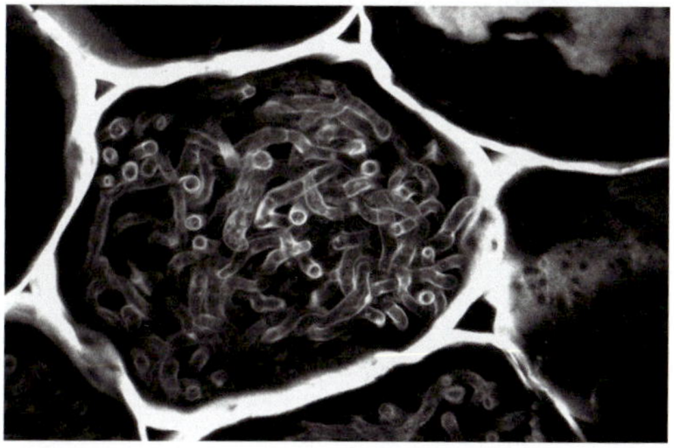

**Figure 2.14** – A cortical cell of a mycorrhizal orchid showing a typical peloton of fungal mycelium; this will be phagocytically assimilated and the root cell will retain its integrity.
Photo: Larry Peterson et al., CNRC

**Figure 2.15** – The monotrope is a parasitic achlorophyllus plant that obtains its carbon source from a tree with which it shares a common ectomycorrhizal fungus.
Photo: Maurice Lalonde, Université Laval

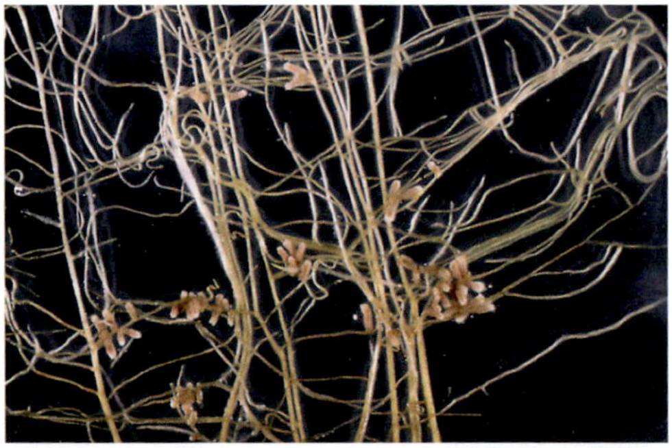

**Figure 2.16** – In legumes, the root nodules caused by the bacteria *Rhizobium* exhibit little or no ramifications and are frequently short-lived.
Photo: Danielle Prévost, Agriculture Canada

Finally, another achlorophyllus genus *Monotropa*, that is not an orchid but rather a member of the Ericaceae, forms an association with one of the ectomycorrhizal fungi associated with the forest trees under which it grows. Under these conditions, the monotrope succeeds in reversing the normal flow of sugar to obtain carbon and energy from the tree, via the fungal mycelium (Figure 2.15).

## *Rhizobium* and *Frankia*

Even today, people still confuse mycorrhizas and nitrogen fixing nodules; however, the latter are structures formed by bacteria, such as *Rhizobium* spp. (bacteriorhizas), or Actinomycetes, such as *Frankia* spp. (actinorhizas). As these structures are not formed by fungi, they are not mycorrhizas (Table 2.1).

Bacteria within the genus *Rhizobium* form symbioses with a large proportion of the extensive and diverse legumes. Although not all species of legumes form nodules with *Rhizobium*, in those that do, the bacteria colonize the roots by penetrating the root hairs, and then move to the root cortex where they provoke the formation of nodules. These nodules (Figure 2.16) become connected to the vascular system of the plant, which ensures nutrition for the bacteria. In this type of association, there is no mycelium or other structure that radiates out into the adjacent soil.

Actinomycetes within the genus *Frankia*, also form nodules on the roots a limited number of vascular plants such as members of the Betulaceae (e.g., Alders) (Figure 2.17), the Rosaceae (e.g., *Dryas*) and the Myricaceae (e.g., Bog Myrtle).

The interest in and ecological importance of root nodules formed by *Rhizobium* spp. and *Frankia* spp., stems from the fact that these symbiotic associations fix nitrogen from the air. Nitrogen is an essential constituent of all living tissue; however, available nitrogen in the soil is rapidly returned to the atmosphere by different denitrifying components of the soil microflora. The nodules formed by *Rhizobium* spp. and *Frankia* spp. actively bring nitrogen back into the ecosystem, making it directly available to the plants that form these associations. Furthermore, this nitrogen then becomes indirectly available to other plants, and animals. Although these symbioses are essential in the plant world, they should not be confused with the mycorrhizas, which have completely different physiological functions and ecological roles. The vast majority of plants that form root nodules also form mycorrhizas. For example, clover and beans that form nodules with *Rhizobium* spp. also have arbuscular mycorrhizal fungi in their roots, and Bog Myrtle that forms root nodules with *Frankia* spp. also forms ericoid mycorrhizas. Actinorhizal *Alnus* (Alder) plants go one step further and host both arbuscular mycorrhizal and ectomycorrhizal fungi in their roots. In contrast, lupin plants, which form nitrogen-fixing nodules with species of *Rhizobium*, are non-mycorrhizal.

**Figure 2.17** – The nodules formed by *Frankia* spp. on the roots of woody species are perennial and highly branched.
Photos: J. André Fortin, Université Laval

## Fungi of the Sebacinaceae

A group of Basidiomycetes that were discovered in India have been shown to have symbiotic properties, and these have recently been added to the list of mycorrhizal fungi. The species *Piriformospora indica* belongs to the Sebacinaceae. The fact that this mycorrhizal fungus is able to grow in pure culture and to enhance plant growth, gives it huge commercial potential. The fungi in this group can colonise roots of a great number of plant species, including members of the Cruciferae, and certain species can form ectomycorrhizas. The research into this group of 'mycorrhizal' fungi is still in its infancy, but *P. indica* is likely to be of great interest over the next decade. The availability of commercial inocula would allow the potential agricultural and horticultural value of these fungi to be tested.

## A brief look at symbioses in the animal kingdom

There are a great many symbioses in the animal kingdom. Take for example *Hydra*, which owes its green colouration to an alga that lives within it and which provides it with a source of energy, just as in the corals (see Chapter 1). *Cyanophora paradoxa* is a ciliated protozoan, which contains a large chloroplast. When the protozoan divides the chloroplast does likewise. Examination using an electron microscope shows that the chloroplast is in fact a cyanobacterium that possesses its own nucleus. Nevertheless, during the course of this symbiotic coexistence, the cyanobacterium has lost an important part of its DNA and it is now completely reliant on its protozoan host for survival. The only DNA that it has retained is the part necessary for the synthesis and functioning of chlorophyll. This is similar to the chloroplastic DNA that we find in green plants today. Spinach leaves owe their green colouration and photosynthetic capacity to a symbiotic alga that was housed in the cells of an ancestral organism.

There are many more example of symbioses in the animal kingdom, take for example the spectacular phenomenon that occurs on the beaches near the marine biology station of Roscoff in Brittany (France). At ebb tide, the sandy beach gradually becomes green. If you walk along the beach, the sand returns to its normal colour almost immediately. Closer examination reveals the presence of a minute green worm, *Convoluta roscoffensis* (Figure 2.18). In its larval stage, this worm absorbs certain single-celled algae and, by a complex mechanism, integrates these into its tissue. By so doing,

**Figure 2.18** – *Convoluta roscoffensis* is an example of a symbiotic association between an animal and an alga. Larvae of this species ingest algal cells. These find their way into the tissue of the worm and the symbiotic worm becomes autotrophic. However, it must expose itself to sunlight in order to benefit from this association.
Photo: Xavier Bailey, Station de biologie marine de Roscoff

the worm becomes autotrophic and no longer needs to feed. However, it must come to the surface of the sand to takes advantage of the sun's energy and the sugars produced by the chloroplasts in the algal cells.

More incredible still, are those molluscs and annelid worms that live close to the fumaroles on the floor of the Pacific Ocean. The sunlight does not penetrate to these depths, and so photosynthesis is impossible. However, the high concentration of $H_2S$ in the water has allowed these organisms to establish close symbiotic associations with chemo-**autotrophic** bacteria that are able to obtain energy by oxidising $H_2S$. This energy source resembles that used in batteries.

The stomach floras of ruminants, and the gut floras of termites and certain other insects, are also amazing. Termites break down the greatest quantity of plant tissue in the world. They are able to digest cellulose by associating with a multitude of micro-organisms that live within their digestive tract. Among these organisms is the protozoan *Mixotricha paradoxa* (Figure 2.19). This species is equipped with four long flagella, which are used in locomotion. The surface of this protozoan is covered by a regular helical arrangement of protuberances. At the base of each protuberance is a cellulolytic bacterium and at the tip there is a cilium. Microscopic examination shows that each cilium is in fact a spirochete bacterium. The latter has a synchronous undulating movement that agitates the intestinal fluid (chyme), ensuring a constant supply of cellulose. In short, life in symbiosis can be quite astounding.

## FIGURE 2.19 – *MIXOTRICHA PARADOXA*

*Mixotricha paradoxa* lives in the intestine of termites, where it helps digest cellulose. This protozoan possesses flagella, and the surface of the cuticle is covered with protuberances. At the base of each protuberance is a cellulolytic bacterium and at the tip is an undulating cilium. Closer observation using an electron microscope shows that the cilia are in fact spirochete bacteria. The constant movement of the latter ensures the movement of chyme providing a constant supply of cellulose for the cellulolytic bacteria.

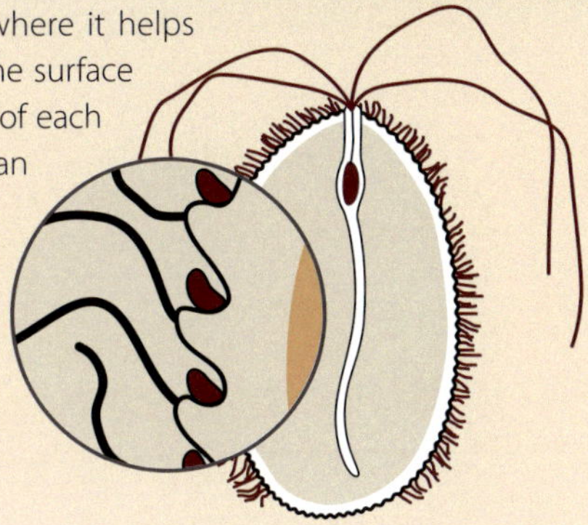

CHAPTER 3

# The origin of plant symbioses and their role in the evolution of life on earth

It is important to understand that plants play a fundamental role in the creation, regeneration and maintenance of the thin layer of soil in which they grow, and on which all life on earth depends.

One of the most common examples typically used to illustrate how marine life probably gave rise to life on dry land, is that of certain fish that gradually evolved into amphibians, developing a double breathing system and transforming their fins into limbs.

What is less commonly mentioned is how plants evolved, and how they succeeded in colonising dry land, where, little by little, they transformed rock and rock fragments into a fertile soil. These rudimentary soils subsequently allowed the establishment of other organisms on dry land.

Before elementary soils could develop, primitive marine organisms had to emerge from the water and establish themselves on bare rock. Natural turnover of these organisms, which included algae, fungi and bacteria, provided the nutrient rich organic matter necessary for the development of soil.

## The appearance and role of lichens

More than 450 million years ago, the appearance of freshly emerged volcanic rock provided a suitable environment that was to favour the evolution of lichens. Prior to this, the earth had been entirely covered by water. A large number of marine organisms had access to this new environment and probably competed for niches. Under such circumstances, it is easy to imagine a continual succession of aquatic organisms on a given substrate as one species out competed another. Some organisms almost certainly found food to ingest, or may have parasitized other organisms. Up to this point, nothing differs from the usual model of evolution. However, in an extraordinarily important turning point in the evolutionary process, symbiotic associations started to play a key role in the development of life on dry land. Rather than favouring the most competitive organisms, natural selection favoured those organisms that were able to form symbiotic associations.

Certain algae and marine fungi, after having exhausted all the possible means of competition and parasitism, found that life in symbiosis offered a remarkable solution for survival on rock outcrops. These organisms formed an association that was particularly fruitful and which led to the first terrestrial flora, the lichens.

In Chapter 2, we saw that the photosynthetic algae in lichens produce organic molecules for both symbiotic partners using carbon dioxide and energy from the sun. In return, the fungal partner provided access to mineral salts present in cracks within the rock. This new and symbiotic association was highly successful, and for several million years, lichens were the only form of plant life on dry land. With time, they succeeded in colonizing not only the coastline (Figure 3.1), but also rock outcrops further inland. The lichens developed the capacity to live in some of the most hostile of environments, absorbing the mineral nutrients necessary for growth directly from rain water and from the rock substrate on which they grew. Following alternating wet and dry periods, the expansion and contraction of lichens developing in cracks on the rock surface, led to physically weathering of the rock and the formation of finer mineral particles. Furthermore, the lichens produced organic acids that contributed to chemical weathering of the rock surface, enabling them to extract mineral nutrients. In addition, as several species of lichens were able to use atmospheric nitrogen, when they died, their remains provided organic matter rich in minerals and nitrogen. This organic matter would have been slowly mixed with the inorganic mineral particles, such as gravels, sands, silts, and clays, present at the site. Thus the rock outcrops started to accumulate the basic ingredients necessary for the development of that loose crumbly mix of organic and inorganic fractions that we know as soil.

**Figure 3.1** – Lichens were the first symbiotic organisms to colonize rock outcrops.
Photo: J. André Fortin, Université Laval

## The appearance and role of arbuscular mycorrhizas in the evolution of vascular plants

Well after the appearance of lichens, a second successful attempt by plants to colonize dry land occurred, this time on the alluvial deposits. These deposits were principally comprised of fine loose material gouged from the rock surface by glacial action and other processes of erosion, including the physical and chemical action of lichens. Run-off, streams and rivers transported this material down slope, where it accumulated in alluvial plains that were to provide a favourable environment for the second successful establishment of plant life on dry land (Figure 3.2a), which would lead to the rise of the vascular plants.

As had been the case with lichens, this second colonization was due to a mutualistic association, but this time it was the arbuscular mycorrhizal symbiosis. Investigations using molecular tools have shown that the first arbuscular mycorrhizas appeared in the Devonian, approximately 450 million years ago. This has also been confirmed by the observation of structures in 400-million-year-old fossils of *Rhynia* that closely resemble the characteristic arbuscules found in the roots of arbuscular mycorrhizal plants today. Structures typical of this mycorrhizal association have also been found in other plant fossils obtained from a wide range of geological formations.

Natural selection for this symbiosis played a determining role in the success of vascular plants. It allowed their survival, proliferation and expansion, and the subsequent evolution of hundreds of thousands of new species. Indeed, the evolution of land plants should in fact be considered a case of co-evolution. Today, the majority of vascular plants still live in symbiosis with arbuscular mycorrhizal fungi, which are all grouped within the Glomeromycota. Mosses, ferns, certain gymnosperms (conifers) and the majority of angiosperms (flowering plants) are hosts to this group of mycorrhizal fungi. This symbiosis not only plays a role in mineral nutrition, it also enhances the host plant's tolerance to drought, disease-causing organisms and other environmental stresses. Interestingly, although arbuscular mycorrhizal plants are widely distributed across the surface of the planet, they still favour alluvial soils, sands, silts and clays.

Once again, this provides a clear example of natural selection favouring a mutualistic approach. The fact that the arbuscular mycorrhizal symbiosis has played such a fundamental role in the evolution of vascular plant species, underlines how life in symbiosis can be a powerful driver of the evolution process. Indeed, it is possibly more powerful than competition, parasitism or predation.

## The appearance and role of ectomycorrhizas in the evolution of gymnosperms

In the Carboniferous period (approximately 300 million years ago), prevailing environmental conditions and the appearance of a novel plant molecule, lignin, favoured the development of much larger vascular plants, such as the tree ferns. These colonized all the available alluvial sites and dead plant tissue provided huge quantities of woody debris. The presence of this abundant substrate proved important for saprophytic fungi, and it allowed an important wave of speciation. It was at this time that the Basidiomycetes appeared, an important group of fungi with the enzymatic capacity to breakdown lignin and cellulose.

Alluvial material only made up a relatively small proportion of the available substrates on the emerged continents. Further inland, were massive rock outcrops that were sometimes covered with glacial **moraine** consisting of coarse gravel and large rocks, some of which were undoubtedly covered with lichens. These upland or mountain ecosystems were constantly subjected to leaching of mineral nutrients by rain and melt water. As these upland systems were unable to retain these nutrients, they were washed down to the alluvial plains in the lowlands. The arbuscular mycorrhizal plants that had evolved on the latter sites were limited in their ability to grow on these upland sites, and it was the higher fungi, the Ascomycetes and Basidiomycetes, that were to provide the key to tackling the problems associated with colonizing this environment.

With time, a number of plants evolved away from the arbuscular mycorrhizal symbiosis towards a completely new association with certain Ascomycetes and Basidiomycetes, the same types of higher fungi that are omnipresent in forest soils today. This split from the arbuscular mycorrhizal association first occurred in some of the gymnosperms (conifers). For example, the Pinaceae (e.g., *Abies*, *Larix*, *Picea* and *Pinus*) adopted the ectomycorrhizal symbiosis, while the Taxaceae (e.g., *Chamaecyparis*, *Taxus* and *Thuja*) have conserved the more ancestral arbuscular mycorrhizal symbiosis. Even today, almost all ectomycorrhizal plants are woody species and include certain angiosperms (e.g., *Betula*, *Fagus* and *Populus*). Reminders of this evolutionary pathway exist, especially in tropical regions, where pine roots may exhibit low levels of arbuscular mycorrhizal fungal colonization. Moreover, several tree species, such as the poplars, regularly form both arbuscular mycorrhizas and ectomycorrhizas.

Members of the Pinaceae and several other ectomycorrhizal species were particularly successful in colonizing the coarser substrates characteristic of upland and mountainous environments (Figure 3.2b), and the moraines left behind by retreating glaciers. The success of ectomycorrhizal plants in these environments was facilitated by three factors: the formation of phenol- and tannin-rich organic matter that is largely resistant to bacterial decomposition, which leads to the accumulation of organic matter on the soil surface; the acidification of the substrate by the activity of saprophytic fungi; and the almost exclusive and efficient means by which ectomycorrhizal fungi can extract nitrogen from organic materials for the benefit of their hosts. Rainfall is generally higher on mountain summits; however, the ectomycorrhizal association has reduced the leaching of mineral nutrients towards the plains, and has allowed the establishment and maintenance of soil fertility in these environments.

## The evolution and role of ericoid mycorrhizas

Ectomycorrhizal fungi are extremely efficient in extracting nitrogen from organic matter, and this contributes to soil acidification and the development of a raw **humus** (see also **mull**, **moder** and **mor**) or peat layer on the surface of the mineral soil. The accumulation of humus and the process of substrate acidification are then further

**Figure 3.2** – Present day examples showing three stages in the evolution of land plant ecosystems. (a) An alluvial plain dominated by arbuscular mycorrhizal plants; the coniferous forest in the background is ectomycorrhizal and typical of boreal or mountainous regions. (b) Evolution of the ectomycorrhizal symbiosis allowed gymnosperms, such as fir and spruce, to colonise upland and mountain environments. (c) Ombrotrophic sphagnum moss peatlands are often dominated by ericaceous plants that form ericoid mycorrhizas. Note the stunted spruce tree in the foreground.

Photos: a: J. André Fortin, Université Laval; b: Fôret Montmorency, Université Laval; c: Josée Landry, Université Laval accelerated by the retention of water, and the release of phenols and tannins from decomposing plant material. Taken to the extreme, this may lead to the formation of peatlands (Figure 3.2c). Once again, it was a new root symbiosis, the ericoid mycorrhizas, which allowed ericaceous plants to survive and grow in this rather hostile environment. The capacity of their mycorrhizal fungi to mobilise nitrogen, further acidify the substrate and tolerate heavy metals, effectively halted the decomposition of organic matter. In the majority of cases, only mosses are able to develop on the surface of such substrates, and over the years this has led to the build up of important of quantities of **peat**, which is highly valued by plant growers.

## The evolution and role of orchid mycorrhizas

Although in evolutionary terms orchids are a fairly recent family, the biology of orchid mycorrhizas is highly complex. The processes of evolution and natural selection have led to an incredible diversity within this family of plants, and to do them justice we would need to devote a whole book to this group. However, the number of scientists studying orchid mycorrhizas is relatively small. This is perhaps because, generally speaking, the orchids that are typically cultivated seem to exhibit little or no dependence on mycorrhizal fungi once fully grown. However, this is not the case for the vast majority of the 30,000 species of orchids found in the wild. Because of the current limited economic importance of most of these species, little financial support is available to study the mycorrhizal associations they form. Nevertheless, this is likely to be an important field of research in the future.

## Breaking free from mycorrhizas

Although diverse mycorrhizal fungi have accompanied vascular plants since their first appearance on dry land, and have probably played a key role in enhancing plant speciation, certain groups of plants no longer form mycorrhizal associations. Examples of such plants include members of the Cruciferae (e.g., cabbage) and the Chenopodiaceae (e.g., beetroot and spinach). In the legumes, plants such as lupins are also non-mycorrhizal; however, they have developed **proteoid roots** to increase nutrient uptake.

## The evolution of nitrogen fixing symbioses

Compared to the arbuscular mycorrhizas that are 450 million years old and ectomycorrhiza that are 250 million years old, the nitrogen fixing symbioses, such as those involving *Rhizobium* and *Frankia*, are relatively recent in origin, dating back 80 and 60 million years, respectively.

CHAPTER 4

# The physiology of mycorrhizas

Regardless of the type of mycorrhizal association, there are six general ways in which plant functions are enhanced by the presence of mycorrhizas. These are the absorption of mineral nutrients, the absorption of water, hormonal activity, protection against pathogens, resistance to diverse environmental stresses, and soil aggregation.

## Absorption of mineral nutrients

The first function that was attributed to mycorrhizas was the enhanced absorption of mineral nutrients, notably those that are poorly mobile in soil, such as phosphorus and zinc. The increased ability of mycorrhizal plants to absorb mineral nutrients from soil is principally due to the fact that the fungal mycelium associated with mycorrhizal plants greatly increases the surface area in contact with the soil solution. It is estimated that the surface area of roots under a square metre of prairie soil is about 1 $m^2$; however, the surface area of the arbuscular mycorrhizal fungal mycelium associated with these roots may be up to 90 $m^2$. Non-mycorrhizal plants do not have to develop extensive root systems to search for mobile mineral nutrients such as potassium and nitrogen. However, nutrients such as phosphorus and zinc, which are rapidly fixed onto substances within the soil matrix, are not mobile. This leads to the development of depletion zones around the root. In the latter case, non-mycorrhizal plants must invest energy into root growth to search for necessary nutrients. The energy invested in developing roots is energy that is no longer available for aerial growth, and reproduction. Mycorrhizal plants overcome this problem by investing in the development of a network of hyphae in the rhizosphere. Indeed, mycorrhizal fungi, particularly arbuscular mycorrhizal fungi, are able to produce a huge absorptive surface with only a limited investment of resources by the plant. Thus, heavily colonized mycorrhizal plants are able to invest more energy in the construction of aerial tissues, such as stems, leaves, flowers and fruit.

Although the enhanced nutrition of mycorrhizal plants is largely linked to an increase in the surface area that is in direct contact with the soil, recent studies have clearly shown that the mycelium of certain ectomycorrhizal fungi is able to breakdown organic matter. This is achieved either alone or in synergy with other soil micro-organisms and it allows direct access to mineral elements trapped within plant and animal remains (Figure 4.1). This mechanism enables ectomycorrhizal fungi to have a positive effect on the nitrogen nutrition of plants. Furthermore, it has recently been shown that like some of the fungi in lichens, certain ectomycorrhizal fungi are able to extract potassium from rocks and make it immediately

**Figure 4.1** – The extraradical phase of an ectomycorrhizal fungus associated with birch (a) in three soils: (c) shows the mycelium developing in soil from a birch forest, and (b) and (d) shows the extraradical mycelium colonizing the remains of soil nematodes.

Photo: David Read et al., *Canadian Journal of Botany*

**Figure 4.2** – The extraradical phase of an ectomycorrhizal fungus showing a greater preference for potassium-rich feldspath (K) than for pure silica quartz (Q).

Photo: Ann Rosling, *New Phytologist*

available to their hosts (Figure 4.2). Recent research has shown that the hyphae of arbuscular mycorrhizal fungi are closely associated with soil bacteria, and that a structural relationship develops between the different organisms, resulting in the formation of a bacterial biofilm (Figure 4.3).

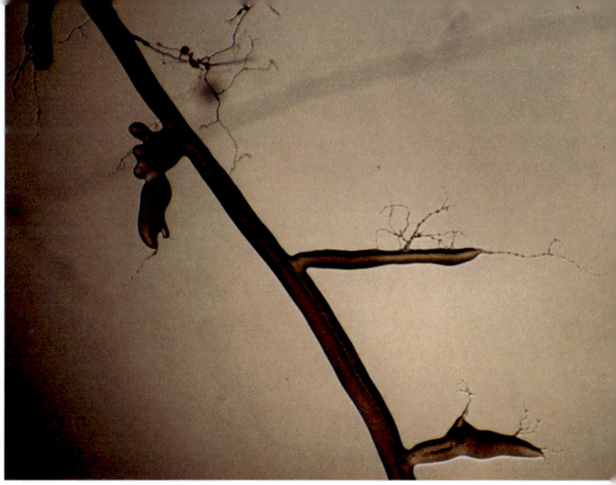

**Figure 4.3** – A bacterial biofilm almost completely covering the extraradical mycelium of an arbuscular mycorrhizal fungus grown *in vitro*.
Photo: J. André Fortin, Université Laval

These biofilms are an integral part of the mycorrhizosphere and have diverse functions. Notably, they have been shown to help dissolve minerals, particularly phosphorus, from the soil matrix, making certain nutrients more readily available for both the fungus and its host plant.

In legumes that are simultaneously colonized by nitrogen-fixing *Rhizobium* and arbuscular mycorrhizal fungi, there is a synergistic and positive effect on plant growth.

## Water absorption

The increased contact between mycorrhizal plants and the soil that leads to an improved absorption of mineral nutrients also results in increased water absorption. Therefore, plants that are colonized by arbuscular mycorrhizal fungi have a greater resistance to drought stress than non-mycorrhizal control plants. This is because the fine fungal mycelium associated with the former, is able to extract water from cracks and pores within the soil matrix that are not accessible to plants roots. When maize plants are subjected to different levels of drought stress, mycorrhizal plants show a greater growth rate than non-mycorrhizal plants, and this difference increases with increased drought stress (Figure 4.4). This may be due to the added fact that under drought stress, the mycorrhizal fungus is also able to produce a signal that ensures rapid closure of stomata in host plant leaves, which potentially helps prevent permanent wilting.

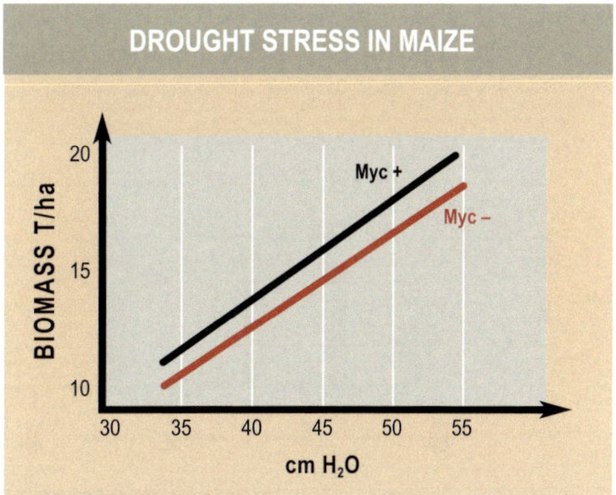


**Figure 4.4** – Mycorrhizal maize plants are better able to withstand drought stress than non-mycorrhizal control plants. The difference in the increase in biomass is greatest when water availability is at its lowest.

Interestingly, while mycorrhizal fungal colonization of wheat grown in fertilized soil tends to have little affect on mineral nutrition, the fungus does enhance the water use efficiency of its host. This may improve the drought stress resistance of mycorrhizal plants, which, in turn, may limit premature flowering and seed production, and so increase yield.

## Hormonal activities

Ectomycorrhizal fungal associations, which typically only form on woody plants, produce growth regulators, such as auxins, which alter root morphology. It only requires very low concentrations of these hormones to produce visually obvious modifications to short roots. For example, in members of the Pinaceae, the ectomycorrhizal short roots show hypertrophy of the cortex and a characteristic dichotomous branching – this is particularly noticeable in the genus *Pinus*. Although there is typically no dichotomous branching in some of the other genera within the Pinaceae, such as Abies, the alterations to root morphology that are induced by the fungus are still clearly visible (Figure 2.10a). Generally speaking, the degree of ramification is proportional to the quantity of auxins that the fungus liberates into the root tissue (Figure 4.5). Although this hormone production increases the root surface, it is also believed to increase the transfer of sugars from the root to the fungus. Overall, the hormones produced by the fungus affect the general appearance of the plant, favouring the growth of aerial parts over that of roots. In other words, as we saw earlier, the fungus provides the plant with a low cost alternative to the heavy investment required for root production.

In the arbuscular mycorrhizal association, at least two plant hormones, jasmonic acid and ethylene, are involved in the colonization process. These molecules also cause biochemical modifications in the rest of the plant. These hormonal modifications even intervene in the mechanisms that control fungal colonization of the host plant root, and so ensure that an equilibrium is maintained between the mycorrhizal fungus and its host.

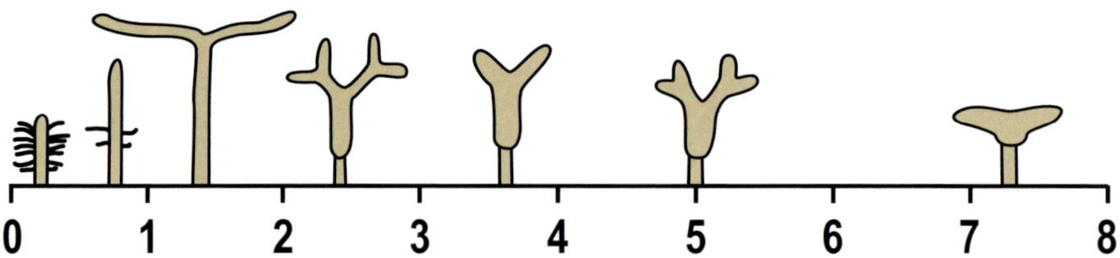


**Figure 4.5** – In the laboratory, the effect of ectomycorrhizal fungi on the development of pine roots can be simulated by treating them with different quantities (0 to 8 mg per litre) of auxin. The morphology of mycorrhizal short roots typically observed in nature corresponds to treatments of between 2 and 5 mg per litre of auxin. In the absence of auxin, the development of small short roots (on the left) is unaltered and they conserve their root hairs.

## Glomalin and soil aggregation

Water stable soil aggregates are fundamentally important for soil fertility as they improve soil aeration, and water and mineral nutrient availability for plants. The stability of these aggregates is enhanced by the presence of mycorrhizal fungal hyphae. The network of mycelium produced by arbuscular mycorrhizal fungi is a dynamic structure that can grow several millimetres per day. To maintain its nutrient and water uptake efficiency, the developing network selectively shuts down certain hyphae. This strategy allows the fungus to concentrate its efforts in nutrient-rich areas. The lifespan of many fine lateral branches of the network is approximately one week. Prior to closing off these sections, the fungal cytoplasm is retracted, leaving behind only the external hyphal wall. These empty tubes serve to reinforce soil aggregates. Furthermore, arbuscular mycorrhizal fungal hyphae also excrete a glycoprotein, glomalin (Figure 4.6a). This protein contributes directly to the pool of stable organic matter in the soil. Indeed, glomalin is highly resistant to decomposition and accumulates within the soil matrix. Sarah Wright, who was the first to discover this molecule, considers that this substance is more abundant than humus in many soils, and that it represents one third of all soil-sequestered carbon. In some soils, mycorrhizal fungi may form 40% of the microbial biomass and up to 30% of the root biomass. They can therefore produce important quantities of glomalin. Several studies have shown that this molecule plays a key role in the structural stability of soils, where it acts as glue, binding fine soil particles together into aggregates (Figure 4.6b). As well as reinforcing

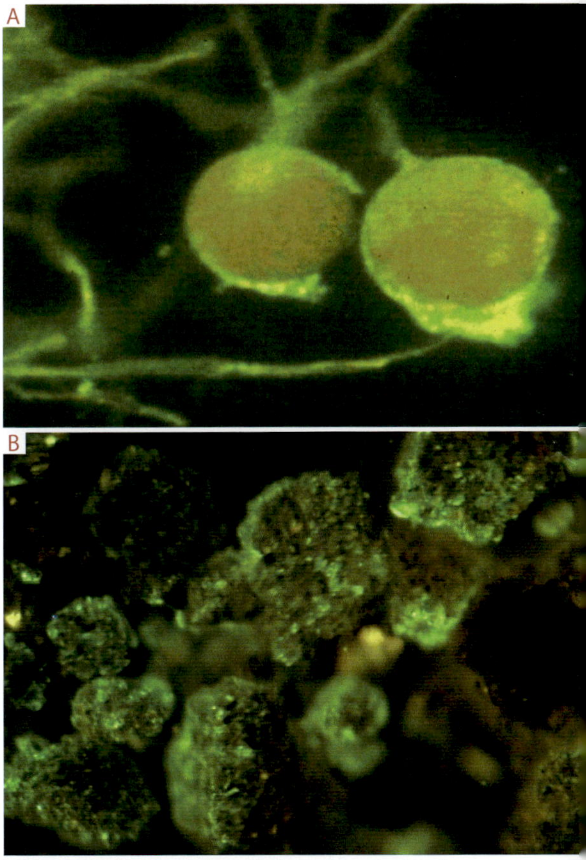

**Figure 4.6** – A specific antibody has been developed for glomalin, the glycoprotein produced by arbuscular mycorrhizal fungi. When this antibody is associated with a fluorescent marker (in this case pale green) it can be used to locate glomalin, either on the surface of the mycelium or spores (a), or on and in soil aggregates (b).
Photos: Kristin Nichols, USDA, ARS

soil aggregates, dead mycorrhizal fungal mycelium also contributes to the pool of soil organic matter, which is also an important factor in the cohesion of soil aggregates. Therefore, to conclude, the mycelial networks produced by mycorrhizal fungi contribute physically, biologically and chemically to the formation and maintenance of stable soil aggregates.

**Figure 4.7** – The plants shown were all inoculated with a species of *Pythium*, a pathogenic fungus. Those in the bottom half of the photograph were also inoculated with the arbuscular mycorrhizal fungus *Glomus intraradices*. The latter plants clearly show enhanced resistance to the disease.
Photo: Susan Parent, PremierTech

## Protection against pathogens

In nature, plants are continually under attack from bacteria, nematodes and important fungal diseases (Figure 4.7).

Mycorrhizal fungi help to protect host plant roots against attack from pathogenic fungi in two ways, and in two places: the rhizosphere and the root tissue.

The rhizosphere microflora is comprised of all the micro-organisms within the zone of influence of the root.

Micro-organisms within the rhizosphere, and above all the mycorrhizosphere, are confronted with differing degrees of competition and antagonism. This results in the establishment of a diverse and equilibrated microflora. In such environments, pathogenic

### THE RHIZOSPHERE OF MYCORRHIZAL ROOTS

A living root not only modifies its environment by extracting mineral nutrients, it also exudes soluble organic substances. In addition, cells shed from the root surface, and the death and turnover of a certain number of roots, contributes locally to the pool of soil organic matter. As already mentioned, the mycelium of mycorrhizal fungi may constitute an important proportion of the soil biomass, and, like roots, this mycelium can exude soluble organic substances. Moreover, dead hyphae contribute to the pool of soil organic matter. Therefore, when considering mycorrhizal roots, it is appropriate to consider the microflora of the mycorrhizosphere. This comprises all the soil micro-organisms occurring within the zone of influence of the mycorrhizal root.

fungi are often unable to proliferate and their numbers remain relatively low. Experiments conducted on such interactions, show that propagules of the pathogenic fungi that cause *Fusarium* wilt of tomatoes and root rot of carnations, are three to ten times fewer in the rhizosphere of mycorrhizal plants than in the rhizosphere of non-mycorrhizal plants.

### MYCORRHIZOSPHERE AND RHIZOSPHERE

The **mycorrhizosphere** is an extension of the rhizosphere and comprises the zone of influence of the extraradical phase of the mycorrhizal fungi on the soil surrounding it. While the zone of influence of the rhizosphere extends only a few millimetres out from the root, that of the mycorrhizosphere extends for several centimetres.

The **rhizosphere** is defined as the zone of influence of the primary root on the soil that surrounds it. The mineral nutrients and water that are found there are susceptible to be absorbed by the root. The root also exudes chemical substances around it, for example amino acids, which affect microbial populations, including mycorrhizal fungi. Micro-organisms within these populations also influence the development and the physiology of the root and its associated mycorrhizal fungi.

The second mechanism by which mycorrhizal plants may enhance their resistance to disease is through a modification of the physiological activity of the root. When the host plant is attacked by a pathogen, the mycorrhizal fungi may react by producing antibiotics against the organism in question. During the course of evolution, mycorrhizal fungi have developed the ability to produce physiological defence mechanisms, without being negatively affected by them. This mechanism allows mycorrhizal plants to limit the development of pathogens, without limiting the activities of the mycorrhizal fungus itself.

There are a growing number of documented cases of synergies between soil bacteria and mycorrhizal fungi in the control of root diseases.

Furthermore, a number of recent studies have shown that the biochemical modifications induced in the mycorrhizal plant through the action of jasmonic acid, has a negative impact on insect species grazing on the plant's foliage. Therefore, certain arbuscular mycorrhizal fungi help prevent insect **herbivory**.

Finally, the composition of the arbuscular mycorrhizal fungal flora in and around the roots of plants can modify interspecific competition.

## Resistance to environmental stresses

Ectomycorrhizal trees in the boreal forest grow in acid soils that typically reach values as low as pH 3. Without mycorrhizal fungi, the roots of these trees would be unable to grow. In fact, at neutral or slightly alkaline pHs, aluminium, iron and manganese remain relatively insoluble and therefore these elements do not negatively affect plant development. However, under acidic conditions, these metals become highly soluble and their concentrations in the soil solution can reach levels that are toxic to plant survival and growth. Under these conditions, ectomycorrhizal fungi are able to protect the roots of their host trees. One mechanism by which they achieve this, is through the sequestration of potentially toxic metals in molecules known as siderophores.

In peatlands, which can reaches pH 2.5, ericoid mycorrhizal fungi allow their plant hosts to grow, despite the potential toxicity of these environments. However, the exact mechanism involved remains to be elucidated.

The situation is reversed in calcareous soils. For example, the Aleppo pine that grows around the Mediterranean is found in soils exceeding pH 7. Under these conditions, iron, which is essential for plant growth, is insoluble and so unavailable to the roots. In such soils, non-mycorrhizal plants develop iron deficiencies, and the subsequent reduction in the amount of chlorophyll produced, leads to the chlorosis (i.e., a yellowing) of their needles. However, the ectomycorrhizal fungi typically associated with the trees growing in this region, secrete organic acids (e.g., oxalic acid) that modify the pH in the vicinity of the host roots. This increases iron mobility and enhances uptake of this element by mycorrhizal plants.

Earlier, we saw that mycorrhizal fungi can enhance the water uptake of their hosts. Studies with *Acacia tortilis*, which grows in the Sahara, show that mycorrhizal plants have a much greater tolerance to drought stress.

The surface temperature of slag heaps can become excessively high, which negatively affects root growth and mycorrhizal activity of plants used for site remediation. Nevertheless, the ectomycorrhizal fungus *Pisolithus tinctorius* can tolerate temperatures of nearly 40 °C in the laboratory. This fungus has been successfully used to inoculate pines growing in hot regions, and on slag heaps in Pennsylvania (USA).

In summary, the formation of mycorrhizal associations modifies several aspects of the physiology of the host plant. It is this wide range of advantages that has allowed the development, the conservation and the diversification of these symbiotic associations over the course of evolution. Natural selection has encouraged the universal distribution of mycorrhizal plants in all the ecosystems of the world. In contrast, natural selection constantly rejects parasitic organisms that contribute no benefit to their hosts. These organisms are forced to rely on closely defined, and changing, niches. The only time that these organisms succeeding in occupying an important role in the life of plants, and in shaping ecosystem processes, is during epidemic outbreaks.

The creation of highly artificial environments, such as agricultural monocultures and single-species forestry plantations, offer parasitic and pathogenic organisms the conditions they need to become firmly established, and to cause a great degree of damage. It is often because of these artificial systems, that all fungi become labelled as harmful organisms that must be eliminated; however, this is far from true.

In the past, symbioses have been considered to be on the borderline of disease, suggesting that to understand a symbiosis, we need to understand pathological processes. However, we are of the opinion that it is better to consider parasites as failed symbioses, which have not been accepted by their hosts. Furthermore, these negative associations have been constantly rejected by natural selection.

CHAPTER 5

# Mycorrhizas in ecosystems

## The role of symbioses in the evolution of terrestrial ecosystems

Whether we take a prairie, a tropical rainforest, a temperate deciduous forest, a coniferous forest, or a peatland, the presence of mycorrhizal fungi is never really very obvious. Up until the end of the last century, plant morphologists and anatomists typically studied root tissue produced from seeds germinated on filter paper, or the roots of plants grown using soilless culture techniques. One of the main reasons behind this was to reduce the damage caused by soil particles to microtome blades during the preparation of thin sections. This led to the general belief that roots growing in nature developed and functioned as those observed in the laboratory. As a result, it has only been quite recently that plant physiologists and ecologists have started to appreciate the importance of mycorrhizal symbioses in the colonization, evolution, functioning and even the distribution of terrestrial ecosystems.

We have already highlighted the fact that the appearance of the arbuscular mycorrhizal symbiosis about 450 million years ago played a determinant role in the establishment of plant life on dry land. In addition, the co-evolution of plants and mycorrhizal fungi had an undeniable impact on the diversification of vascular plants. Furthermore, the arbuscular mycorrhizal symbiosis was a determining factor in the development of soil and the evolution of terrestrial ecosystems.

Before looking at some of the ecosystems present today, it is worth briefly summarizing the information presented in the previous four chapters concerning the evolution of the land surface after the emergence of the continents. Briefly, the freshly emerged landmasses were literally rock outcrops, and lichens were the first to successfully colonize these inhospitable environments, a niche that some species still exploit today. Under the combined influences of diverse erosion and weathering processes (e.g., glaciers, freeze-thaw action, precipitation, and chemical weathering by acids produced by lichens), rock outcrops were progressively altered and broken down into smaller and smaller particles. This resulted in the formation of gravels, sands, silts and clays. Under the influence of rain, the smallest particles were transported down slope, where they eventually collected in loose deposits. Furthermore, the soluble minerals that would serve as nutrients for terrestrial plants were also washed down slope, continually enriching the alluvial deposits. It was precisely in this environment that the first symbiotic vascular plants developed, and from which they began to colonize the continents. This combination of factors, including the establishment of plants on dry land, led to an increase in the build up of dead organic matter, which provided a food source for saprophytic bacteria, fungi, protozoa, annelids and arthropods. With time, the microflora and fauna, and mesofauna incorporated this organic matter into the underlying mineral substrate of sands, silts and clays. This led to the development of the first soils and of highly fertile **clay-humus combinations**, which similar organisms still help maintain today. It is also important to remember that certain mineral nutrients are highly soluble (e.g., nitrogen, calcium and potassium), and others, much less so (e.g., phosphorus and zinc). The establishment of mycorrhizal associations is particularly important for plant uptake of nutrients in this latter group.

## Distribution of the different types of mycorrhizal symbioses in nature

In the following section, we use an example from the St. Lawrence River watershed (Québec, Canada) (Figure 5.1), to illustrate how the dominant mycorrhizal type changes with altitude as we move from alluvial plains to mountain summits.

### Hickory- and basswood-sugar maple forests

On the fine deposits left behind after the disappearance of the Champlain Sea, there is a sequence of sugar maple (*Acer saccharum*) forests that change in composition with increased altitude. This sequence starts with the hickory-sugar maple forest, which is a relatively rich ecosystem with approximately 130 species of vascular plants, and where hickory is typically the only ectomycorrhizal species. This ecosystem is also home to one or two non-mycorrhizal species. All the other species growing on this melanic brunisol with a mull-type humus, form arbuscular mycorrhizas. Higher up the slope, there is a transition towards the basswood-sugar maple forest ecosystem. The soil on which this forest type grows is a sombric brunisol. This

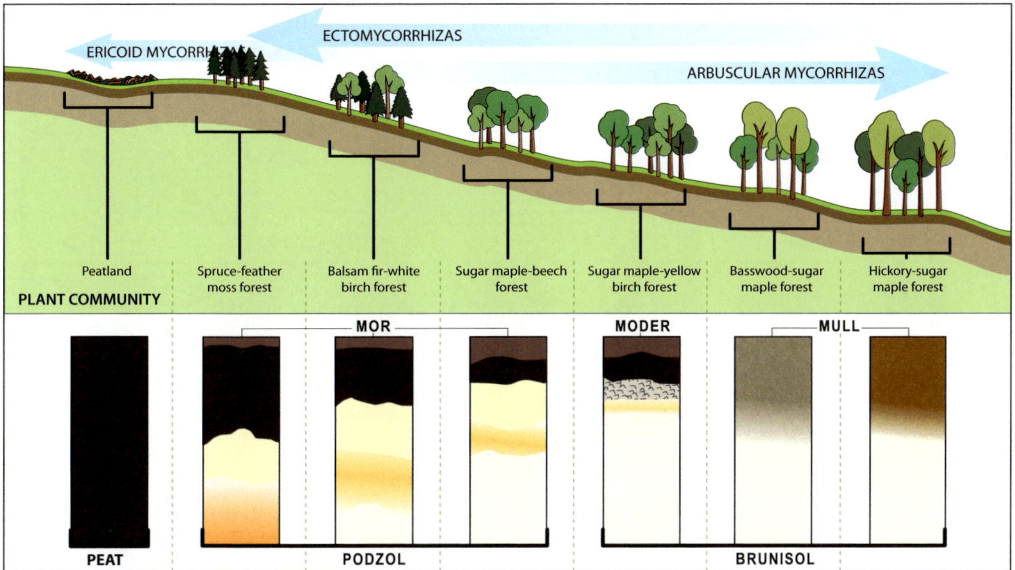


| PLANT COMMUNITY | Peatland | Spruce-feather moss forest | Balsam fir-white birch forest | Sugar maple-beech forest | Sugar maple-yellow birch forest | Basswood-sugar maple forest | Hickory-sugar maple forest |
|---|---|---|---|---|---|---|---|
| Ericoid mycorrhizas | ± | ± | – | – | – | – | – |
| Ectomycorrhizas | +++ | +++ | ++ | + | ± | ± | ± |
| Arbuscular mycorrhizas | – | ± | ± | ++ | +++ | +++ | +++ |
| pH | 2.9 | 3.6 | 3.9 | 4.2 | 5.8 | 6.5 | |
| C/N | 37 | 27 | 28 | 26 | 14 | 14 | |
| %N | 1.5 | 1.8 | 1.1 | 1.1 | 0.41 | 0.59 | |
| Number of plant species | 16 SPP | 20–25 | 67 | 76 | 91 | 100 | |
| 50% turnover rate (years) | 45.3 | 34.7 | 4.6 | (2.3) | 2.8 | 1.7 | |
| Earthworms | – | – | – | ± | + | + | |

**Figure 5.1** – A transect through the valley of the St. Lawrence to the summits of the Laurentian Hills, showing the transition from one mycorrhizal type to another with increased altitude. Arbuscular mycorrhizal dominated deciduous forests occupy the alluvial soils at low altitude (today these have been largely replaced by agriculture) and ectomycorrhizal coniferous forests occupy the coarser soils at higher altitude.

Source: J. André Fortin, Université Laval.

system is also dominated by the arbuscular mycorrhizal symbiosis; however, there are typically only about 100 species of vascular plants present. In these two more alkaline brunisols, earthworms are abundant and draw organic matter deep into the soil, favouring the formation of clay-humus combinations that enhance soil fertility. The decomposition of organic matter in these soils is extremely rapid and the nitrogen cycle is complete (Figure 5.2), which provides plants with nitrates. In these two forest ecosystems, leaf litter is absent from the soil surface during summer. These are soils that are particularly well suited to agriculture and the majority of the plants cultivated in them are, as elsewhere, arbuscular mycorrhizal.

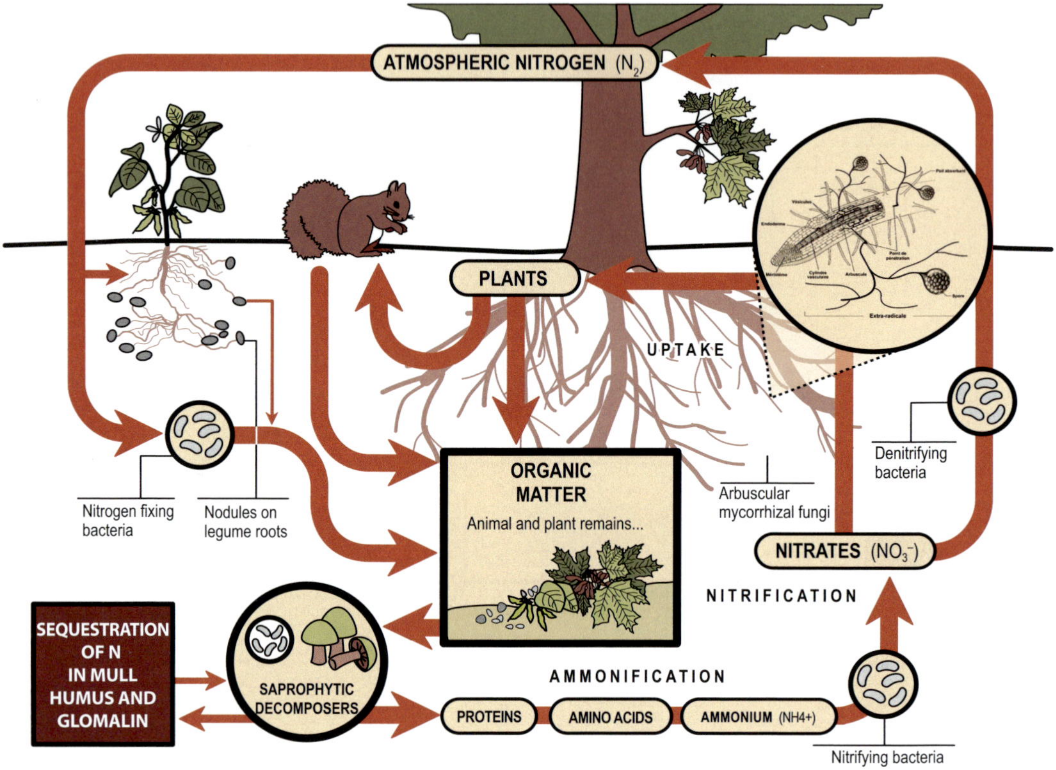


**Figure 5.2** – The nitrogen cycle in arbuscular mycorrhizal-dominated broad-leaved deciduous forests. Saprophytic decomposers and micro-organisms release nitrogen from animal and plant remains. The amino acids released from proteins are rapidly transformed into ammonium and nitrates. The latter is the principal form of nitrogen used by arbuscular mycorrhizal plants. A similar nitrogen cycle occurs in prairie and agricultural soils.

Source: Modified from Wikipedia by J. André Fortin, Université Laval

## Sugar maple-yellow birch and sugar maple-beech forest

On the moraines found in the foothills, the basswood-sugar maple forest gives way to sugar maple-yellow birch and sugar maple-beech forests. Yellow birch and American beech are ectomycorrhizal, and both are relatively abundant within their respective forest ecosystems. When compared with the maple forests at lower altitudes, the number of species of vascular plants in much reduced. Although the soils in these systems are still classed as brunisols, they are more heavily leached,

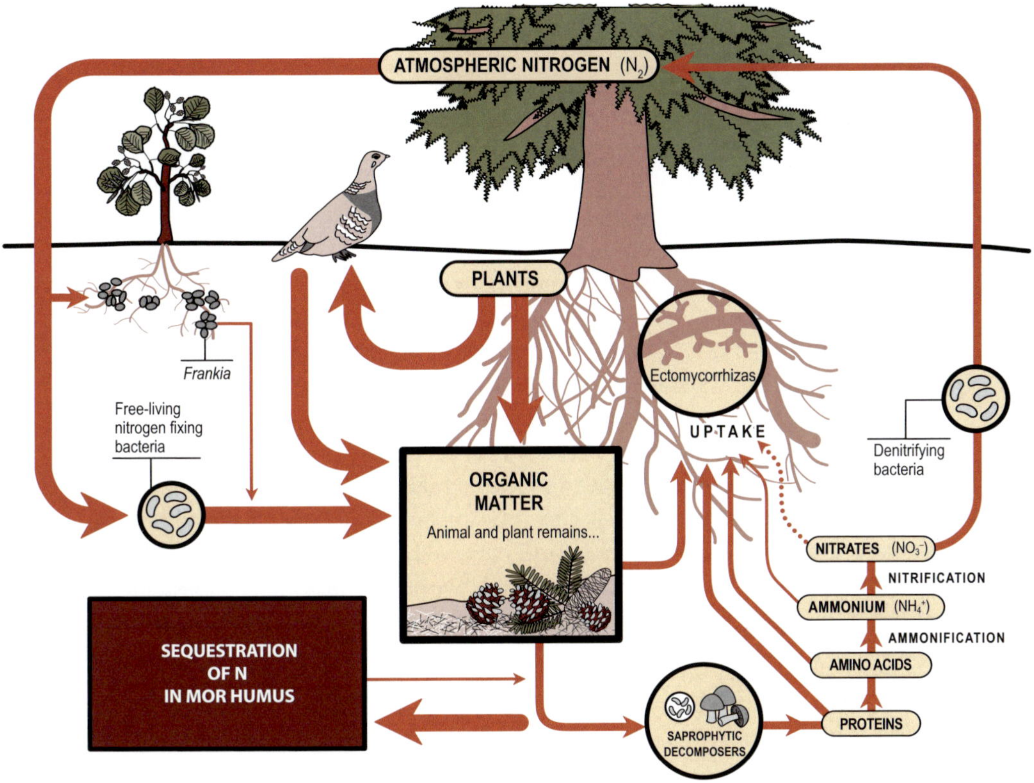


**Figure 5.3** – The nitrogen cycle in ectomycorrhizal coniferous forests. The activity of saprophytic decomposers is limited and ectomycorrhizal fungi rapidly colonize animal and plant remains, and directly extract amino acids. Those amino acids that are lost are transformed into ammonium, which ectomycorrhizal fungi can easily absorb. The nitrification rate of any remaining ammonium is low, and the nitrates formed are used by the few arbuscular mycorrhizal plants that are able to live in these environments (e.g., ferns and raspberry).

Source: Inspired from Figure 5.2, J. André Fortin, Université Laval

due to increased precipitation at higher altitudes. The soil pH is slightly acidic and earthworms are rare or absent. To a certain degree, soil arthropods, mostly insect larvae, mix the organic mater into the mineral soil, giving a moder-type humus. The decomposition of organic matter is slower in these systems and the nitrogen cycle tends to be only partial, with ammonium being the dominant nitrogen form present. These systems are characterized by the presence of a layer of organic matter that is clearly distinct from the underlying mineral soil.

## Balsam fir forest

The amount of precipitation is positively correlated with altitude, and the underlying **parent material** is often made up of coarser moraines. Conifers dominate the forests on such sites, with yellow and then white birch co-occurring with balsam fir. The number of arbuscular mycorrhizal plant species is markedly lower; however, raspberry and several species of ferns may be abundant in certain more open areas. There is also a transition from brunisols to podzols, the latter being characterised by a thick layer of acidic (approximately pH 4), only slightly decomposed, organic matter or raw humus. Earthworms are absent from these soils and the activity of insect larvae is no longer sufficient to incorporate the organic matter into the underlying soil. The water that percolates from the humus layer is acidic and leaches the underlying mineral substrate, forming a white layer or horizon, comprised of acid washed minerals. Because the parent material deeper down in the soil profile is less acidic (approximately pH 5), when the leached minerals, including soluble organic matter, come into contact with it, they precipitate out of solution, rather like in a chromatograph. This produces a horizon, generally below the rooting zone, that is rich in iron deposits and organic matter. In these soils the process of nitrification (the oxidation of nitrogen to give nitrates) is inexistent, and the presence of ammonium, virtually zero. Later, we will see how trees growing in such environments obtain their nitrogen (Figure 5.3).

## Spruce forest

Higher still on the slope, precipitation increases further and the temperature decreases. These sites are occupied by black spruce forest, which is characterized by a very limited range of plant species, and only one or two species form arbuscular mycorrhizas. Mosses are abundant and there are a number of plants that form ericoid mycorrhizas. The degree of podzolization (see above) is even greater and the pH of the humus layer may be as low as pH 3. Because the humus layer is much thicker, the leaching of the surface mineral horizons is even more pronounced.

## Peatland

Finally, peatlands develop in the more poorly drained areas, and sphagnum mosses and ericaceous plants dominate these ecosystems. These sites are characterized by an almost complete breakdown in the decomposition of organic matter, which leads to an accumulation of peat.

## Mycorrhizas and ecosystem dynamics

In summary, arbuscular mycorrhizal trees and other plants are inclined to grow on fine, often alluvial, soils. These tend to be on sites where precipitation is relatively lower, and where the pH is neutral or only slightly acidic. These ecosystems have a rich diversity of plants and a highly active soil fauna. The nitrogen cycle is complete, and includes nitrogen fixation by both free-living and symbiotic bacteria, denitrification, ammonification and nitrification (Figure 5.2).

In contrast, at the opposite end of the scale, ectomycorrhizal trees grow on coarser soils subjected to higher levels of precipitation, and the number of co-occurring plant species is low. The soils are acidic to highly acidic and the activity of the soil fauna is much reduced. The nitrogen cycle is almost completely short-circuited, and the nitrification process is almost completely suppressed and the ammonification process much reduced (Figure 5.3). The blocking of the nitrogen cycle is primarily due to the efficiency of the ectomycorrhizal fungi, which use their extensive extraradical phase to rapidly colonize animal and plant remains. These hyphae quickly remove nitrogen from organic sources before other soil organisms (i.e., saprophytic bacteria and fungi) can access it. Furthermore, ectomycorrhizal fungi also take up phosphorus from organic molecules such as the phytates. Although the nutrition of trees in these forests is largely organic, research has shown that certain ectomycorrhizal fungi, just like the fungi in lichens, produce organic acids (e.g., oxalic acid), which allow them to chemically weather rock and extract mineral nutrients such as potassium and phosphorus (Figure 4.2).

The strong enzymatic capacity of ectomycorrhizal and ericoid mycorrhizal fungi, and their ability to rapidly extract key mineral nutrients from available substrates, are largely responsible for the process of humification in podzols and peatlands. This process effectively stops the leaching of mineral nutrients down slope towards the plains. Therefore, humification plays a vital role in the establishment of forests on thin soils in upland and mountain areas.

It is only relatively recently that the scientific community has started studying the ecology of mycorrhizas, and it has taken a long time for plant ecologists to integrate this dimension into their observations and interpretations of ecological processes and patterns. Many of the initial ecological studies on these symbiotic associations were done in the Northern Hemisphere, and in particular in forest ecosystems.

To enhance our understanding of the role of mycorrhizal fungi on a global scale, it will be important for mycorrhizal research groups throughout the world to take a closer look at the ecosystems in their regions. This will provide a better appreciation of the role of local processes and mycorrhizal symbioses in soil development, and allow the elaboration of an explanation similar to that proposed above for north-eastern North America.

The understanding of the role of mycorrhizal fungi in the establishment, development and stability of terrestrial ecosystems is still in its infancy. Ecologists are just starting to give mycorrhizal symbioses the importance that they deserve. However, it is important to highlight the fact that the methods available for locating, identifying and quantifying the different fungal species within the soil, in particular the arbuscular mycorrhizal fungi, remain difficult and time-consuming. Nevertheless, new molecular techniques are opening interesting avenues for research. Some of these, notably real time polymerase chain reactions, will soon allow the identification of hundreds of soil micro-organisms in a matter of hours.

Given the fundamental role of mycorrhizas in the functioning and stability of natural ecosystems, as well as in those managed by man, it seems evident that research into these symbiotic associations will see important advances in coming years.

CHAPTER 6

# The biology of edible forest mushrooms

By definition, fungi lack chlorophyll and are unable to photosynthesize; instead, these **heterotrophic** organisms, obtaining their energy and carbon from existing organic matter. There are three principal life strategies used by fungi: saprophytic, parasitic and symbiotic. The saprophytic strategy is the most primitive and the most common life strategy found among the fungi. Examples of fungi using this strategy include the Cultivated Mushroom and the Oyster Mushroom. Saprophytes obtain their carbon and energy from the remains of plants, fungi, animals or even micro-organisms. The mycelium of saprophytic fungi penetrates their food substrate, where, in the company of other components of the microbial fauna and flora, they breakdown the organic matter and extract nutrients, so ensuring the recycling of minerals. The parasitic strategy is a more evolved and specialised one. Examples of parasitic fungi include those that cause the heart rot of trees, and rust infections in wheat. These fungi penetrate the tissues of a living organism to obtain some or all of the carbon and energy that they need for growth. Certain parasitic fungi cause only limited damage (e.g., fungi causing leaf spots), while others may kill their host. Examples of the latter type include the fungi involved in chestnut blight and Dutch elm disease. Both of these fungi were accidentally introduced from Europe, and have practically eliminated their North American plant hosts. Although pathogenic fungi, even those that kill their hosts, have diversified and become specialized during the course of evolution, they have not become as widespread as the symbiotic fungi that form mycorrhizal associations. Parasitic fungi could almost be considered as fungi that attempted to form symbioses, but failed. Symbiotic fungi have the most evolved life strategy. These fungi obtain at least an important part of their carbon and energy from their host, without causing it any harm. In fact, on the contrary, these fungi protect their hosts against pathogens and environmental stresses, and improve their nutrient status. A large proportion of edible forest mushrooms are the fruitbodies of symbiotic fungi. Among these are species of Chanterelles, Boletes and Truffles; however, deadly poisonous species of Amanitas are also symbiotic.

## Reproductive structures of some edible species

Although there are several species of saprophytic fungi that produce edible forest mushrooms, the majority of species that are harvested form ectomycorrhizal associations with trees.

Here are a few examples of edible ectomycorrhizal fungal species that, with the exception of the Morels, must be symbiotically associated with a tree to survive and produce their reproductive fruitbodies.

The Hedgehog Mushroom (Figure 6.1a) is always a pleasure to find in the forest. As is the King Boletus, another highly sought after species (Figure 6.1b). This elegant mushroom can reach a considerable size, but it is always preferable to harvest the fruitbodies in the button stage, because as they mature insect larvae rapidly colonize them. The fruiting period of this species is relatively short.

The Chanterelle has an edible fungal fruitbody with numerous qualities (Figure 6.1c) and the discovery of a colony always brings a special thrill. This species is pleasing in both shape and colour, and its unique fragrance means that it can be easily identified, even with your eyes closed. While the majority of fungal fruitbodies have a relatively short lifespan, that of the Chanterelle is about 45 days and, under particularly favourable conditions, may be as long as 60 days. In addition, this species is usually free of insect larvae, it keeps well, and fresh individuals are relatively well suited to long-distance transportation. Furthermore, the Chanterelle has the ultimate quality of site fidelity. Indeed, when you find a colony of Chanterelles at a certain site one year, if the climatic conditions are favourable, you will find them in exactly the same place the following year.

The extraordinary colour of the Orange Cap Boletus (Figure 6.1d) makes this species very easy to find. The Matsutake (Figure 6.1e), which has a strong and distinctive fragrance, is particularly appreciated in Asia and, increasingly so, elsewhere.

This species is essentially associated with pines and develops in late fall. Interestingly, molecular studies have shown that the fungi commonly grouped under the name Matsutake, come from a number of closely related species of varying gastronomic interest.

Morels (Figure 6.1f) are spring fruiting fungi that can be quite difficult to find, and it may take a considerable amount of time to locate a colony. There are at least two species of Morels in the boreal forest of North America. The first is associated with poplar, most often on alluvial soils, and the second with coniferous forests (e.g., pine and spruce). Fruitbodies of the latter, only appear in any great number in the spring following a forest fire. In certain years, large areas of the boreal forest are burned naturally and it is sometimes possible to harvest tens of tonnes of Morels from these sites. However, the mycorrhizal status of Morels is still under debate. This is because the hyphae of certain species are able to form typical ectomycorrhizal structures, but they are also able to grow and fruit in the absence of any mycorrhizal association.

## Growth and development of fruitbodies

Although ectomycorrhizal fungi are unable to decompose cellulose and lignin, they are very efficient at extracting nutrients from diverse mineral and organic sources. Although the symbiotic state remains essential for completion of the fungal lifecycle, recent studies have shown that a number of mycorrhizal species exhibit various degrees of saprophytic activity.

Certain ectomycorrhizal fungal species rely solely on their hosts for carbon and energy. These include the Laccarias and the Hebelomas. These groups mainly contain species that produce small fruitbodies. However, it is possible that these fungi also possess some saprophytic activity. In contrast, there are a number of species for which the symbiotic relation is essential, but which are capable of an important degree of saprophytic activity. This group includes the King Boletus and the Matsutake, and the saprophytic activity of these fungi may be necessary to ensure the development of the large fruitbodies that they typically form. Finally, some other species, such as the Morels, are facultative symbionts.

Different species of ectomycorrhizal fungi occupy different horizons within the forest soil. Chanterelles and Laccarias form their mycorrhizas in the upper soil horizons, which makes them more vulnerable to trampling. In contrast, species of Tricholomas (e.g., the Matsutake) develop their mycorrhizas deeper in the soil (>20 cm).

**Figure 6.1** – Most edible forest mushrooms live in mycorrhizal symbiosis with trees. These include (a) the Hedgehog Mushroom, (b) the King Boletus, (c) the Chanterelle, (d) the Orange Cap Boletus and (e) the Matsutake. Morels (f) occasionally form ectomycorrhizal associations, but their fruitbodies are usually formed in the absence of a symbiotic relationship.

Photos: a, b, c and d: Marie-France Gévry, UQAR; e: J. André Fortin, Université Laval; f: Renée Dumouchel, FauneNord

D

F

E

The fruitbodies of certain species develop over a relatively long period of time. This is true of the Black Truffle, which forms a primordium during the summer that develops into a fruitbody and reaches maturity during the following winter. Although much quicker, the Chanterelle still takes five to six weeks to reach maturity, which, in turn, is much longer than the 10 days that the King Boletus takes to produce its fruitbody.

It is interesting to consider why the majority of ectomycorrhizal fungal species in the coniferous boreal forest only form fruitbodies in late summer or early fall. Evidence suggests that fruitbody production in many species is linked to changes in day length or photoperiod. The reason for this becomes clearer if we consider the pattern of tree growth in the boreal region, which is divided into three distinct stages. At the onset of summer, rapid growth occurs at the top of the tree, and the majority of the products of photosynthesis are mobilized to achieve an increase in height. From the middle of July, increase in height ceases and a dormant apical bud forms, which will only develop the following spring. From this period on, the apex of the tree no longer requires sugars and these are channelled towards the lower parts of the aerial system. This process continues until the middle of August, when growth of lateral branches also stops; it is only then that the third stage, root growth, can start. Clear evidence shows that the sugars produced during late summer and early fall are transported towards the roots, where they also favour growth and function of ectomycorrhizal fungi.

While this is the case in the boreal forest, the season of fruitbody production may be very different in temperate and tropical regions. However, although the mechanism responsible for the differences in periodicity in these systems is unknown, it too is likely to be linked to the transfer of sugars to the host plant root system.

If we return to the boreal forest, it seems that the first factor necessary for the development of mycorrhizal fungal fruitbodies, is an increase in the amounts of sugar reaching the roots. As we have seen, this is triggered by a shortening of the photoperiod that causes aerial growth of the host to cease. Nevertheless, if sufficient water is lacking, the fungus cannot take advantage of this abundant food source for fruitbody production.

It is easy to appreciate that a tree may actively photosynthesize on warm sunny days in late summer and early fall, and that as the aerial parts of the tree are unable to use the sugars produced, that these are transported to the roots where they may stimulate fungal fruitbody production.

On the other hand, on cold overcast days, there will be less sugar arriving in the roots and fewer mushrooms. As overcast days are usually indicative of rain, it seems likely that a period of sunny days interspersed with a few days of rain provides the ideal conditions for an abundant production of mushrooms. However, this model is a simplified one and does not explain all the factors underlying fruitbody production. Indeed, this model cannot be used to explain the case of the Chanterelle, which starts its development in mid-July.

While on the subject of Chanterelles, contrary to what we might at first think, the fruitbodies growing at the same site, at least in the boreal forest, do not necessarily all come from the same colony (i.e., they are not like apples on an apple tree), but are rather a group of individuals that may be genetically distinct. At the base of each Chanterelle there is a limited number of mycorrhizas occupying an area of about 10 $cm^2$. Knowing that each Chanterelle lives on average seven weeks, it is possible to imagine a situation whereby the fungus slowly invests small amounts of carbon in the growth of the fruitbody over the whole of this period, and that it is not reliant on the surplus sugars that arrive in the roots towards mid-August.

It is also possible to imagine that ectomycorrhizal fungi that produce large mushrooms, such as the King Boletus, have a much more important number of mycorrhizas spread over a much greater area (perhaps 1 $m^2$). This would ensure the rapid transfer of sugars over the eight to ten-day-period during which its fruitbody develops.

Soil temperature is also an important factor in triggering fruitbody production in certain ectomycorrhizal species. The results of a long-term monitoring program conducted in a King Boletus experimental production station (the *Boletière expérimentale de la Bosque de Biscarosse*) in southwest France, show a clear relationship between soil temperature, rainfall and fruitbody production. It seems that when the soil temperature crosses the limit of 15 °C (either by a rise or a fall) after or just before a heavy rainfall event, production of fruitbodies occurs in the 14 days that follow. In the case of the Matsutake that only appears in late fall, fruitbody production is likely triggered by temperatures below a certain limit.

It is possible to simulate some of these environmental conditions in the laboratory. For example, pine plants colonized by *Laccaria bicolor* may by grown under long day conditions (18 hours of light) for several weeks to mimic summer in the boreal forest. A sudden decrease in day length to nine hours simulates fall and leads to the development of the fruitbodies of the associated ectomycorrhizal fungus. Under these conditions, the dormant apical bud of the plant forms in less than a week, as if it were preparing for winter. It is then possible to increase the photoperiod back to 18 hours, without breaking dormancy. As long as the plant does not receive a cold treatment or a high level of nitrogen fertilizer, this state of dormancy can be maintained at room temperature for at least 15 months. Such plants can literally become mushroom factories and the combined weight of the numerous fruitbodies produced may greatly exceed that of the plant.

Unfortunately, *Laccaria bicolor* produces a small mushroom that lacks gastronomic interest; it remains to be seen whether a similar approach can be used to produce more interesting mushrooms, such as Chanterelles, Boletes (see below) and Milk-caps.

## Production of trees colonized by edible ectomycorrhizal fungi

For scientific purposes, we can grow tree seedlings in transparent sachets containing a blotting paper support and a nutrient solution. This technique allows us to observe the formation and the development of ectomycorrhizas (Figure 6.2a). In certain species, such as the Laccarias and Hebelomas, we can even obtain fruitbodies (Figure 6.2b). However, for certain edible fungi, such as the King Boletus (Figure 6.2c), only relatively few strains form mycorrhizas in these sachets. Nevertheless, it is also possible to obtain fruitbodies in soil that has been pre-sterilized, or in artificial peat-based substrates (Figure 6.2d). For the production of larger species, such as Boletes (Figures 6.2e and 6.2f), it is important to colonize several seedlings with the same fungal strain; these can then collectively feed the developing fruitbody.

It is impossible to cultivate ectomycorrhizal fungal fruitbodies without the contribution of a host; however, it is possible to obtain the fruitbodies of edible species by producing and planting trees carrying a particular fungal species. This was first achieved in France over 30 years ago, when hazelnut and oaks trees were inoculated with *Tuber melanosporum* and used to produce Black Truffles. Similar tree seedlings are now sold in a number of nurseries, and numerous plantations in different parts of France have since produced truffles. The value of this approach was confirmed when a scientist from New-Zealand, who had noted that soils both locally and in Tasmania shared several properties with those of Black Truffle plantations in southern France, imported trees carrying ectomycorrhizas of *T. melanosporum*.

**Figure 6.2** – Production of ectomycorrhizal fungal fruitbodies under controlled conditions. To test the capacity of a given fungal culture to form ectomycorrhizas, they can be inoculated onto host plant seedlings grown in germination sachets (a) containing an absorbent paper support and an appropriate fertilizer. Certain fungi, such as *Laccaria bicolor*, even form fruitbodies under these conditions. Out of 25 King Boletus cultures tested under these conditions, only one produced ectomycorrhizas with spruce (c). In artificial substrates, it is possible to obtain fruitbody formation (d) by controlling the photoperiod. This method also works with species of *Suillus* (e, f).

Photos: a, b, c and d: Christian Godbout, Université Laval; e and f: J. André Fortin, Université Laval

This was a success, and there are now several economically viable Black Truffle plantations in Tasmania. Today, several attempts are being made to introduce and produce a wide range of Truffles in different parts of the world.

What has proved possible to achieve with the Black Truffle, one of the most capricious of all ectomycorrhizal fungi, should be possible to do with other species, such as the Chanterelle, the King Boletus, other Boletes and the Milk-caps. The King Boletus is currently the focus of such studies in several countries. Unfortunately, efforts made to establish Chanterelle plantations in Sweden have so far failed. Nevertheless, trials conducted in France have shown the potential interest of developing plantations for the Saffron Milk-cap. At present, scientists in Japan and China are very active in this field of research, and in the near future it will likely be possible to control fruitbody production of a wider range of edible ectomycorrhizal fungal species. However, this is unlikely to replace the pleasure of harvesting edible forest mushrooms in the wild.

CHAPTER 7

# Harvesting of edible forest mushrooms

We have already seen that the majority of mushrooms growing on the forest floor live in ectomycorrhizal symbiosis with the surrounding trees. These mushrooms come in a truly remarkable array of shapes, colours and fragrances, and a number of species are of great gastronomic or medicinal value. Forest mushrooms have long been of interest, and they have provided the material for professional scientific investigations, for studies by amateur naturalists, and for chefs. More recently, they have triggered investigations into the possibility of commercial harvesting and mycosylviculture.

## Scientific mycological studies

There are thousands of species of higher fungi, and the physiology and taxonomy (i.e., the naming and scientific classification of species) of these is of interest to a great number of scientists. Mycology is a demanding field of research requiring sophisticated laboratory techniques and equipment, including those suitable for conducting molecular level investigations. Taxonomic work and subsequent discussions between experts often lead to revisions of existing classifications. This means that a given mushroom may end up with several names depending on the authors consulted, and certain species can be particularly confusing. Fortunately, the majority of species do not cause problems. Furthermore, recent breakthroughs in molecular techniques provide rapid answers to certain long-standing taxonomic problems. Taxonomic work requires the development of reference collections in herbaria (i.e., dried specimens) and live cultures. However, scientist are not only interested in naming fungi, they are also interested in studying their biology and ecology. Scientists are currently actively investigating the environmental factors that ensure the development and survival of different species. They are also assessing the potential impacts of commercial harvesting, pollution, climate change, and timber extraction on fungal distribution. Some of these finding will no doubt be applied to the new and growing practice of mycosylviculture.

Given the vital role of fungi in ensuring life on earth, and their involvement in our everyday lives, a greater importance and visibility should be given to mycology in universities throughout the world.

## The amateur mycologist

The growing interest in wild mushrooms in an increasing number of countries has lead to the formation of numerous amateurs associations and clubs. Mycologists joining such groups may be seeking a scientific pastime, or want to learn more about gastronomic or poisonous species.

If you want to learn how to identify wild mushrooms, it goes without saying that the best way is to join your local mycology club. Although there are numerous books on fungal identification, it is difficult for these to replace the popular science courses given by mycology clubs, and the valuable discussions that it is possible to have with other amateurs. Most clubs give a range of courses for beginners and for more advanced amateur mycologists. They also organise outings (Figure 7.1a), hold identification session during the fall, and occasionally organize exhibitions of fresh specimens. Not to mention the fact that these meetings are also excellent social events.

The Internet is an extremely easy way to find your nearest club. If such a group does not exist in your area, you could always organise one yourself. You will probably start with two people and end up with 200!

The work of certain amateur mycologists has led to some remarkable scientific findings. For example, over a 30-year-period, an amateur mycologist, together with a group of friends, collected tens of thousands of data on the fruiting dates of over 300 species of fungi in southern England. Once retired, and with the help of his son who had become a professor in ecology, he complied his data and found that over this period, approximately one third of the species that originally only fruited in the fall, had started to also fruit at the end of winter. Interestingly, the mean February temperature in this area had increased from 3.5 °C in the 1950s, to 5.2 °C.

## Commercial harvesting

In a great many countries throughout the world, the commercial harvesting of edible forest mushrooms has been done for hundreds if not thousands of years. Globally, the annual trade in edible forest mushrooms is currently valued at about 2.5 billion dollars. The majority of this trade involves Chanterelles (Figures 7.1b, e). Truffles, Matsutake, King Boletus, and other Boletes are among some of the other most highly sought after ectomycorrhizal species. In China, close to 100 species are consumed, and in Mexico, it is common to find several tens of species in public markets. In France, Italy and many other European countries, wild mushrooms have been harvested on a large scale for a very long time, and these countries are important importers of edible forest mushrooms. Today, there is hardly a single country where the commercial harvesting of edible forest mushrooms is not practised at some scale.

Currently in the boreal forest of Canada, enormous quantities of edible mushrooms remain unharvested. This is not due to problems associated with the identification of the species present, but rather to logistical problems linked with harvesting and marketing the available resource. The Morel offers a particular interesting case. Gastronomically speaking, this mushroom occupies an almost mythical place. The fruit bodies of some Morels, which are probably facultative ectomycorrhizal fungi, grow in the vicinity of trees such as poplars, and produce fruitbodies in the spring. The harvest of these Morels varies depending on weather conditions. However, in the boreal forest of North America there is another species of Morel that appears in huge quantities in the spring following a forest fire. The huge forest fires that burned in Alaska in 2004 resulted in the harvesting of over 65 tonnes of fresh Morels in 2005. We now know that this phenomenon also occurs in north-eastern North America, and in particular in Québec and Labrador (Figures 7.1 c, d).

The quantity of mushrooms harvested in a given region can reach tens of thousands of tonnes. For example, in the Spanish province of Castilla-Leon alone, 9000 harvesters hold permits and each sends on average 100 kg of mushrooms to one of 70 buying stations. To process such a huge quantity of mushrooms requires an efficient logistical setup (Figure 7.1e) and access to a suitable market (Figure 7.1f).

The harvest of forest mushrooms is highly seasonal and involves a growing number of countries. The structure of the market differs from country to country and can become disordered if there is no organization that provides a minimal code

**Figure 7.1** – The harvest of wild mushrooms. (a) A group of amateur mycologists identifying their harvest after an outing to a local forest. (b) Commercially harvested Chanterelles. (c) Field camp for the harvest of Morels following a forest fire. (d) A young amateur mycologist. (e) Cleaning Chanterelles after a cottage industry-scale harvest. (f) Fresh Chanterelles on display at a market.

Photos: a: J. André Fortin, Université Laval; c, d and e: Domaine de la rivière Mistassini; b and f: Emilian Champignon, inc.

of ethics. Presently, there is an increasing amount of legislation aimed at controlling these activities. However, it is important to avoid a monopolization of the resource by one or two large companies that make a profit at the expense of the harvesters.

## Mycophageous animals

Well before man, diverse species of animals learnt to eat certain fungal fruitbodies. For example, species of cervidae, in particular reindeer and caribou, eat large quantities of epigeous fungal fruitbodies, and wild boars are attracted to, dig out, and eat truffles. However, of all the mammals, it is the rodents that consume the largest quantity and diversity of hypogeous fungal fruitbodies (i.e., those formed underground). For example, fungi make up 90% of the diet of certain species of flying squirrel. Interestingly, it was by studying these small mammals that scientists discovered a previously unknown truffle of gastronomic interest. Other such truffles are probably waiting to be discovered in several other regions of the world.

## Mycosylviculture

In addition to planting trees that carry the ectomycorrhizas of fungi such as Truffles, Milk-caps and Boletes, different forestry regimes, in plantations or in natural forest, are likely to affect the quantity and the species composition of the edible ectomycorrhizal mushrooms produced. For example, in eastern North America, many plantations of white spruce, Norwegian spruce and balsam fir produce good quantities of King Boletus. Furthermore, thinning, and the pruning of lower branches, has been shown to increase the quantity of fruitbodies produced.

In cases were the habitat needs of a given species are known (e.g., the Chanterelle), it is also possible to inoculate chosen sites with fungal spores, in the hope that this might succeed in establishing the species in the area.

## Integrated resource management

In Spain, there is a national program in place that groups all the key players involved in the harvest of forest mushrooms in a given region. This is a huge program that seeks to tackle issues ranging from basic capacity building amongst local harvesters, to the harvest, storage, transformation, and local and international marketing of edible forest mushrooms. A similar model could be adapted for use in other countries that have an abundant edible mushroom resource.

CHAPTER 8

# The production of arbuscular mycorrhizal inocula

## Propagules

In nature, arbuscular mycorrhizal fungi use spores, vesicles and colonized root fragments as propagules. These allow the colonization of new host roots when conditions are favourable for both fungal and plant growth. All the species within the Glomeromycetes produce spores on the extraradical mycelium that develops within the soil, often some distance from the roots. In certain species, spores may be produced in loose clusters, or in tighter clusters referred to as sporocarps. Most of the Glomeromycetes, with the exception of members of the Gigasporaceae, form vesicles within the root cortex of their host plants. When an arbuscular mycorrhizal root is severed from a plant or when it dies, the vesicles behave like spores, germinating and colonizing new roots.

Spores produced by members of the genus *Glomus* are large by fungal standards reaching between 50-300 µm in diameter. However, the spores of species within the genera *Gigaspora* and *Scutellospora* can exceed 500-600 µm, and are easily visible to the naked eye.

In general, species in the genus *Glomus* offer several advantages for the production and use of arbuscular mycorrhizal inocula. By and large, they are prolific, and relatively easy to cultivate aseptically on root-organ cultures, and they can be easily multiplied by transplanting cuttings of colonized roots. Although certain species within the genera *Gigaspora*, *Scutellospora*, *Acaulospora* and *Entrophospora* are also able to grow and sporulate in the presence of root-organ cultures, it is impossible to multiply them using excised segments of a colonized root. This means that new spore-based cultures must be started if a greater quantity of material is required. Furthermore, these genera produce relatively low numbers of spores under *in vitro* culture conditions. This chapter outlines the different approaches required to produce commercial inocula of species within these different genera.

## Pot culture production

The pot culture technique (Figure 8.1a) is the oldest and simplest means of producing inoculum. In fact, this technique is nearly always used for isolating new species or isolates of arbuscular mycorrhizal fungi. For a new isolation, all that is needed is a good-sized pot (e.g., 15 cm); a sterilized substrate such as soil or, better still, a non-organic substrate (e.g., montmorillonite, perlite or sand); and a balanced nutrient solution that is compatible with root colonization (< 25 ppm (parts per million) of soluble phosphorus (P)), the development of an extraradical mycelium and sporulation. The material containing the fungus to be isolated is then placed in the pathway of the developing roots of the host plant used.

**Figure 8.1** – The isolation of arbuscular mycorrhizal fungi can be done using pot cultures (a) of a range of sizes; however, larger containers should be used for the production of inoculum (b).

Photos: Christian Plenchette, Université Laval

Generally one of two types of starting material is used to establish the pot culture: spores or root fragments. If spores are to be used, ideally these must first be extracted from the soil. To achieve this, a small amount of soil is washed through a nest of soil sieves using running water. The mesh of the sieves should range from 250 to 50 µm, with two or three intermediate mesh sizes. The material collected on each sieve should then be cleaned by floatation. This allows the spores to be separated out from the lighter dead organic matter and the heavier mineral fraction. This can be done by layering four or five different concentrations of a liquid (e.g., a glycerol or sucrose solution) to produce a density gradient ranging from 1.2 to 1.3. The density of arbuscular mycorrhizal fungal spores is about 1.25, and these will concentrate in the corresponding layer.

There will also be other soil organisms in the band of liquid where the spores are concentrated. The next step in the spore selection process involves taking a drop of the liquid containing the spores and placing it in a Petri dish. The generally spherical spores can be clearly seen under a dissecting microscope. These are often filled with lipid droplets and have a short stalk (i.e., the remnants of the subtending hypha). The different species of fungi present in the sample can be distinguished by their shape, colour, and the presence or not of certain structures or appendages. We can either choose to use a single spore to start a culture, or to group 15 to 20 morphologically identical spores. Before using these spores to initiate pot cultures, they should be treated with a surface disinfectant, to remove bacteria and possible fungal pathogens.

After this, the spores are transferred to pots containing sterile soil or, better still, an inert substrate such as expanded clay. Ideally, the spores should be placed on a piece of nylon about 5 cm from the soil surface – this stops them being washed to the bottom of the pot during watering. A pre-germinated seed of a suitable host plant (e.g., leek) is then planted just above the spores. Instead of using spores, it is also possible to use segments of a mycorrhizal root. Ideally, root fragments containing vesicles should be used, and the presence of these structures can be readily detected in young roots using a dissecting microscope. Once the starter material (i.e., spores or root fragments) has been placed in the pot and the seed or seedling planted, each pot should be watered with an appropriate low phosphates nutrient solution, and placed in a greenhouse or growth chamber.

After 12 to 15 weeks, a few roots should be harvested, stained and observed under the microscope to look for signs of colonization. If the roots are heavily colonized, the contents of the pot, including the roots, can be harvested as inoculum. The roots should be separated from the substrate and if the species colonizing them typically produce vesicles, the roots can be simply washed, cut into segments, partly dried, and used as inoculum. A spore-based inoculum can be produced by sieving the substrate (see above) to separate out the spores. After having checked that the culture is monospecific, it can then be multiplied at a much larger scale.

There are several variations to this technique, but all allow the cultivation of a maximum number of species of arbuscular mycorrhizal fungi. If using this approach, and if it is important to maintain monospecific cultures, care must be taken to prevent cross-contamination between pots during watering. In addition, precautions must be taken to eliminate the risk of contaminating the cultures with plant pathogens. One particularly good technique is the use of sterile, ventilated bags, which ensure that cultures are kept physically isolated from each other.

## Production in containers

The use of larger containers (e.g., 30 × 50 × 50 cm) (Figure 8.1b) to produce inoculum is a simple adaptation of the pot culture method. It allows the production of large volumes of spores and or colonized roots. This approach can be used to produce inoculum to colonize seedlings grown in trays or for the inoculation of field-grown plants. In certain commercial inoculum production plants, this method is used on a very much larger scale. Nevertheless, one of the main problems associated with the use of this approach is the difficulty in controlling pathogenic organisms. However, this method has been used with remarkable success, notably in Cuba.

In the case of monospecific inoculum production, this technique can be semi-automated by placing the containers inside deep basins that are periodically filled with a nutrient solution, using a pump and a timer.

## Field production

In certain cases, it may be appropriate to consider the production of inoculum in the field. The soil used for this should be heat sterilized by covering it with black plastic and exposing it to strong sunlight. Although this large-scale inoculum production technique can be used to inoculate whole fields, it is less appropriate if the goal is to conserve the arbuscular mycorrhizal fungal biodiversity of a given soil.

## Aeroponic production

Growing plants aeroponically involves suspending a plant in a darkened container and feeding the roots by regularly spraying them with a solution of mineral nutrients. If this approach is to be used for arbuscular mycorrhizal fungal inoculum production, the concentration of nutrients must be compatible with the formation of mycorrhizas, and sporulation. As the nutrient solution is continually recycled in these systems, spores can be harvested from the solution. Following surface sterilization, these spores can be used as inoculum. Commercial inoculum production using this and other similar methods, results in a higher quality product than that obtained from pot cultures or containers. However, in contrast, only a limited number of fungal species can presently be grown using this technique.

## Production under sterile laboratory conditions

All the methods outlined above are effective for producing arbuscular mycorrhizal fungal inocula. Nevertheless, they all have a common problem: the ease with which other organisms and even pathogens can contaminate the inoculum source. The culture of whole plants *in vitro*, and thus in the absence of unwanted organisms, can be done, but it is difficult, and needs complex and expensive equipment and facilities. However, in the mid-1980s, a novel sterile technique using root-organ cultures was developed for the production of these fungi.

The easiest way to obtain a root-organ culture is to germinate a seed under sterile conditions. The tip of the radicle is then transferred to a Petri dish containing an appropriate solidified culture medium that has all the nutrients (i.e., sugar, vitamins and mineral salts) needed for root growth. This technique can support

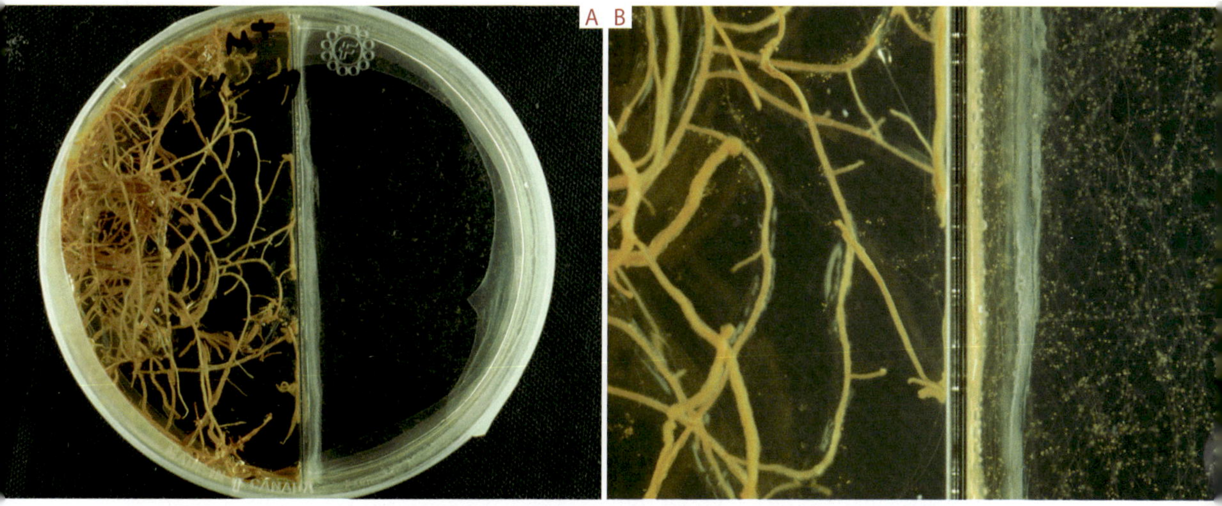


**Figure 8.2** – In (a), the left-hand compartment contains a genetically transformed carrot root-organ culture growing on a medium containing mineral nutrients and sugar; in the right-hand compartment, the medium only contains mineral nutrients and simulates the soil environment. In (b), a similar Petri is shown, but at a higher magnification. On the left, the medium in which the root-organ has developed allows root colonization by an arbuscular mycorrhizal fungus, but it does not favour the extensive development of the extraradical phase or sporulation; on the right, the mycelium has passed over the separation barrier, and has developed an extensive mycelium and produced thousands of spores.

Photos: Marc St-Arnaud, Université de Montréal

the growth of isolated roots of some plants (e.g., tomato) for in indefinite period of time, as long as the root apex is regularly transferred to a Petri dish containing fresh media. Nevertheless, it is easier to produce strong, rapidly growing, root-organs with the help of a simple genetic transformation that occurs naturally in soils. This transformation is produced by the bacterium *Agrobacterium rhizogenes*. Carrot root-organs produced using this approach are particularly vigorous, and are extremely useful for work with arbuscular mycorrhizal fungi. Briefly, the bacterium transfers genetic material to the plant and this subsequently induces extensive root growth. Once this has been done, the bacteria can be removed using antibiotics, and the roots produced will continues to grow. Root-organs provide a particularly valuable tool for producing large quantities of high quality arbuscular mycorrhizal fungal propagules (Figures 8.2a and b).

Unfortunately, at the moment, only a few species of arbuscular mycorrhizal fungi are able to grow and sporulate under these *in vitro* conditions. Certain modifications to the culture medium, for example by adding sterilized soil extracts, could perhaps help overcome this limitation.

## Commercial *in vitro* inoculum production

Although the exact details of commercial inoculum production techniques and inoculum formulation are closely guarded trade secrets, they are based on the laboratory method outlined above. This is adapted for commercial-scale production by simply using larger containers and optimising growth conditions. At present, two companies, one in Canada and one in India, use such techniques to produce massive quantities of inoculum. The commercial inocula that are produced are available in different forms in garden centres, and through horticultural, agricultural and nursery suppliers, and for golf courses and vegetation restoration work. Given the **axenic** nature of the products in question, they can also be used in scientific research, where contaminant-free material is frequently required. (visit the official Internet sites of Teri Industries and Premier Tech for further details, see page 131 for the addresses).

## Formulations

The propagules obtained vary according to the inoculum production method used. Some result in the production of colonized roots (with vesicles), spores and mycelium, while others only produce colonized roots (with vesicles), or spores.

For the formulation of commercial inocula that are mixed with a carrier, the product must be proved to contain neither contaminants nor pathogens. This is difficult to show and only specialized laboratories can do such tests. However, it goes without saying that the inocula produced under sterile conditions provide a better chance of success than those produced under non-sterile conditions. Ideally, the inoculum should be separated from the substrate, and the roots, spores and mycelium dried. The companies using this approach take a number of precautions to assure the quantity and quality of propagules in the inocula they produce. The formulation of this type of inoculum can face problems that are almost impossible to overcome, unless it is produced in an inert substrate (e.g., expanded clay or perlite) and under strictly controlled conditions. As the propagules produced under sterile conditions face fewer problems, they are likely to dominate the inoculum market in the future.

For the species of arbuscular mycorrhizal fungi that form vesicles in host roots, colonized roots can be used as a simple inoculum. Once the roots have been extracted from the substrate, they should be surface sterilized, dried, chopped and stored under cool dry conditions prior to use. Roots produced using aeroponic culture techniques can be used in the same way, and may be ground to separate out vesicles. In the case of propagules produced sterilely on root-organ cultures, disinfection prior to grinding is not needed.

For spore-based inocula, the spores must be extracted from the substrate. Although this is difficult, it can be done by floatation using a liquid of a given density (e.g., a sucrose or glycerol solution, see above) that allows the heavier substrate to sink and the lighter spores to float on the surface.

The formulation phase can be done in different ways and is a challenging stage for the marketing of inocula. Each country has its own regulations that serve to ensure the quality, quantity and efficiency of the marketed product. The inocula currently on the market take a number of forms. These include spore suspensions (mainly for scientific research), semi-liquid gels, wet-able powders, granules, or ready prepared peat-based potting composts. This last option has the advantage of providing an environment that is compatible with, and favourable for, the development and beneficial action of arbuscular mycorrhizas.

CHAPTER 9

# Mycorrhizas in horticulture

Arbuscular mycorrhizal fungal inocula are sold as concentrates or already incorporated into potting compost. Inocula of the latter type were developed to favour the establishment and functioning of the mycorrhizal symbiosis. If a concentrated inoculum is used, it is essential to make sure that the growth medium into which it will be incorporated is compatible with mycorrhizal development. It is particularly important to avoid using substrates containing high levels of phosphorus. It should also be kept in mind that during the course of evolution, the ancestors of certain cultivated plants, such as cabbages, broccoli, beetroot and spinach, lost their ability to form mycorrhizal associations. Therefore, it is important to know whether the plants being grown form mycorrhizas or not, because there is no advantage to be gained trying to inoculate plants that are naturally non-mycorrhizal. If in doubt, the label on the inoculum should list the most common non-mycorrhizal species.

The importance of mycorrhizal formation for plants depends, at least from a nutritional point of view, on the level of phosphate fertilizer in the soil and the presence of suitable propagules (see mycorrhizal dependency, Chapter 10). Figure 9a shows carrots growing in a typical field soil with a medium level of phosphorus (100 ppm (parts per million)). In contrast, Figure 9b shows carrots growing in the same soil after fumigation with a substance that does not affect plant growth, but which eliminates mycorrhizal fungi. Under the latter conditions, carrot plants showed only minimal growth, and to obtain a level of growth similar to that observed in Figure 9a, it was necessary to increase the level of phosphate fertilizers to 400 ppm. A similar pattern was observed when leek plants were tested in the same soils (Figures 9c and 9d). However, when the soil was reinoculated with a suitable arbuscular mycorrhizal

**Figure 9** – (a) Carrot plants in natural soil, with mycorrhizas. (b) Carrot plants in fumigated soil, without mycorrhizas. (c) Leek plants in natural soil, with mycorrhizas. (d) Leek plants in fumigated soil, without mycorrhizas. (e) Leek plants in natural soil, in fumigated soil, and in soil reinoculated with *Glomus intraradices*. (f) Apple saplings inoculated with different species of mycorrhizal fungi; the non-inoculated control is on the left.

Photos: Christian Plenchette, Université Laval

fungus, the soil that had been fumigated allowed a level of plant growth that exceeded that of the natural soil (Figure 9e). This was perhaps due to the suppression of pathogens, the fact that different species of arbuscular mycorrhizal fungi have different effects on a given plant species (Figure 9f), or a combination of both these factors.

## Direct sowing in the garden

A great number of horticultural and market garden species are sown directly into the soil. This can be done manually or using various types of machinery. The type of mycorrhizal inoculum used also varies depending on whether the plants are being grown under small- or large-scale production systems.

The inocula produced for gardeners come in the form of a peat-, clay- or quartz-based granular mix, or in a semi-liquid gel. These inocula can be used to colonize and enhance the growth of crop plants such as carrots, beans, onions (from both seed and sets), and cucumbers. The seedbed can be prepared by digging a small shallow trench and lining this with the chosen inoculum. A limited amount of mineral or organic fertilizer with a minimal amount of phosphorus ($<75$ µg/g of substrate) can then be added in a lateral band. This should be placed in such a way as to avoid contact between the inoculum and the fertilizer.

The seeds or sets are then sown or planted directly onto the inoculum and covered with soil to the depth recommended for the plant in question. The bed is then regularly watered, taking care to limit the amount of fertilizer used. The technique is much the same for potatoes, except for the fact that the inoculum is placed around the eyes on the top of the seed potato, as this is where the roots will develop.

Ready-prepared diluted inocula are becoming increasingly common and they are ready for use in seedbeds. These are basically potting soil mixes of peat, vermiculite and perlite, to which mycorrhizal fungal propagules have been added. These mixes are the same as those used for producing plants in trays and pots. They contain everything necessary for plant growth, including an optimal amount of mycorrhizal fungal propagules. Furthermore, the level of fertilizer they contain is totally compatible with the establishment and development of mycorrhizas. In addition, these substrates provide an ideal environment for seed germination as they maintain an optimal level of humidity. It is difficult not to produce a good crop of vegetables using such products.

For large-scale market gardens, the formulation of the inoculum employed is highly important. The machinery used for sowing includes precision seeders and so the formulation of the inoculum must be compatible with this equipment. It is also important to make sure that a sufficient number of propagules are placed in the vicinity of the seed. This is to ensure that the symbiosis establishes rapidly on the developing root system. The positioning of the inoculum in the row is equally important; it must be placed such that the roots of the plant reach it as early on in the growing season as possible.

## Seedling for transplantation

The most effective means of colonizing flowering plants and crop plants such as tomatoes, peppers and leeks that are grown for transplantation, is to introduce the mycorrhizal inoculum at the same time as the seed is sown. If commercial potting mixes containing mycorrhizal fungal propagules are used, then the process is simple, and colonization rapid. In contrast, if concentrated inocula are used, each container must first be filled to within 2 cm of the top with potting mix, a thin layer of inoculum added, the seed sown and a covering of substrate applied. Using this method, the roots growing through the layer of inoculum are rapidly colonized by the mycorrhizal fungus, which afterwards colonizes the whole of the root system. After a number of weeks, which varies with the plant species being grown, the substrate used and the recommendations of the inoculum manufacturer, it will be necessary to fertilize the plants. However, care should be taken not to add too much soluble phosphorus (<75 µg/g of substrate).

The effect of the mycorrhizal symbiosis will be quicker in commercially prepared substrates, because the mycelial network (like a three dimensional spider's web) will develop rapidly throughout the substrate. In the case of a concentrated inoculum, the mycelium must grow progressively into the substrate.

Commercially prepared substrates and high quality concentrated inocula are free of pathogenic organisms. However, if a fungicide must be used to control a certain fungal disease, it is essential to make sure (i.e., by reading the label) that it is not harmful to mycorrhizal fungi. In many countries, the companies that produce fungicides are not presently obliged to show that their products are harmless to mycorrhizal fungi; nevertheless, this is likely to become a requirement in the near future.

Commercial plant production relies to an increasing extent on multi-celled containers that are filled and seeded automatically. Each compartment typically contains 1-2 $cm^3$ of substrate, although this varies according to the size of the seed used. This is the ideal place to introduce a mycorrhizal inoculum, but the small size of the compartments requires that the inoculum be well mixed to ensure that each cavity receives a sufficient quantity of viable propagules. The seedlings produced in multi-celled containers are then automatically transplanted into larger containers. Later we will discuss why it is advantageous to transplant into a substrate that also contains an evenly distributed inoculum.

## For cuttings

The best way to inoculate cuttings is to proceed in much the same way as for seedlings, making sure that the inoculum is placed close to the site from which the newly formed roots will emerge. The simplest way to achieve this is to insert cuttings into a substrate that already contains mycorrhizal fungal propagules. Some cuttings require a well-drained substrate and if this is the case, sand or perlite can be added to the commercially prepared diluted inoculum.

## For *in vitro* production

*In vitro* propagation has become a commonly used technique for a large number of commercially produced plants. Under such systems, the most appropriate time to inoculate the plantlets is when they are transferred from *in vitro* to normal *ex vitro* conditions. Recently, in an attempt to increase plantlet survival during this phase in plant production, several scientists have managed to produce mycorrhizal plants under sterile multiplication conditions. However, this approach seems to require a $CO_2$-enriched atmosphere (5000 ppm). Nevertheless, plants pre-colonized *in vitro* show an enhanced tolerance to drought stress during acclimation. This breakthrough is of great interest for scientific research, and in the future is likely to be extremely useful in automated commercial-scale production units.

To return to current *in vitro* plant production systems, the best way of using mycorrhizas is to inoculate plantlets at the time of acclimation. This can be done by transferring the plantlets to a substrate containing a suitable inoculum, following the method outlined above for seeds and cuttings. As under normal *in vitro* to *ex vitro* transfer, the evapotranspiration of the plantlets must be strictly controlled.

## Transplantation to pots

Mycorrhizal seedlings obtained from traditional sowing techniques or the use of multi-celled containers, and plants grown from cuttings or plantlets produced *in vitro*, must sooner or later be repotted. However, is there an advantage to using a substrate containing mycorrhizal fungal propagules or not? In theory, the root system of the young plant to be transplanted is already well colonized by the mycorrhizal fungus that was present in the original inoculum, and the substrate should contain an extensive extraradical mycelium.

If this plant is transferred to a pot containing a large volume of substrate without inoculum, it will take time for the mycelial network on the root system to grow into all parts of the new substrate. However, if a well-mixed substrate containing the same fungal strain is used, the propagules will germinate throughout the substrate and their mycelia will fuse to the network already established on the plant being repotted.

The result of transplanting an already mycorrhizal plant into a substrate inoculated with the same mycorrhizal fungus is that the extraradical mycelium (a three dimensional spider's web-like structure, see Chapter 2) and the plant, will more rapidly benefit from all the substrate. This will lead to enhanced plant resistance to drought stress, protection against pathogenic fungi, and access to a greater amount of mineral nutrients.

In conclusion, a pre-inoculated plant, whether it comes from a seed, a cutting, our an *in vitro* produced plantlet, that is planted into a substrate bearing the same mycorrhizal inoculum, will rapidly benefit from all the advantages of the mycorrhizal association.

## Outplanting

Although the importance of mycorrhizas for the majority of commercially important plants has been widely emphasised over the last 30 years, the majority, if not the totality, of plants sold in garden centres, or used in commercial market gardening, are rarely mycorrhizal when purchased.

Earlier chapters in this book highlighted the fact that mycorrhizas are ubiquitous in nature. So why inoculate? One reason is that working the soil greatly reduces the number of active mycorrhizal propagules present. In addition, not all species of mycorrhizal fungi produce the same results for a given plant growing in a given soil. However, commercial inoculum producers are in a position to offer inocula that contain species and/or strains that are capable of colonizing a wide range of plants and of growing in different soils. Furthermore, these selected strains are efficient in terms of nutrient uptake, and typically offer protection against a wide range of environmental stresses.

Plant producers commonly analyse the mineral nutrient content of their soils, and it would be possible to add to this an evaluation of the number of mycorrhizal fungal propagules present. Although such analyses are not yet available, they would allow plant producers to add inoculum to soils containing an insufficient quantity of propagules. At the moment, this type of analysis is only done in research laboratories. However, generally speaking, following soil manipulations, the mycorrhizal fungal inoculum potential is typically insufficient.

It is clear that certain advantages can be gained by adding inoculum to the planting hole when transplanting pot-grown plants. This is particularly true if the plants are being planted in soil that has been heavily worked, and the advantages are the same for both annual plants and perennials (e.g., roses, shrubs and trees). The best approach is to place the inoculum evenly around the clump of soil that has remained attached to the roots.

In the case of shrubs and above all trees, several species form symbioses with ectomycorrhizas. Commercial inoculum producers generally offer gardeners a product that contains both propagules of arbuscular mycorrhizal and ectomycorrhizal fungi. While this approach is suitable for small-scale plantings, plant producers would be better to use a specific arbuscular mycorrhizal or ectomycorrhizal fungal inoculum to suite the species being cultivated.

Plants such as tulips and hyacinths that develop bulbs also form arbuscular mycorrhizas. Many species and varieties must be planted deep in the soil and studies have shown that, except in deciduous forests, mycorrhizal inoculum at these depths is almost absent. The best means of ensuring that horticultural bulbs flower for several years is to place a little mycorrhizal inoculum in the bottom of the planting hole. This practice is completely compatible with the usual addition of bone meal.

Certain plants are sold as bare rooted stock (e.g., roses, peonies, and many other shrubs and trees) and these greatly benefit from the application of a suitable mycorrhizal inoculum when they are being planted. If a concentrated inoculum is being used, a little can be applied directly to the root system, and a further amount can be mixed with the soil dug out of the planting hole. However, it is probably more advantageous to use a commercial peat-based mycorrhizal inoculum. This is because, by using a sufficient quantity, the plant will benefit from both rapid root colonization and an improvement in soil quality.

In the foreseeable future, gardeners and horticulturalists will increasingly request plants that are guaranteed to be mycorrhizal.

## Establishing turf

A lawn can be established either by seeding or laying turf, and mycorrhizal fungi are beneficial in both cases. For a lawn that is already established, it is also possible to apply an inoculum to the surface after aeration. Although the products generally available on the market have been developed for use on golf courses, they can be used in the same way on a smaller scale for domestic purposes.

Inocula for use during seeding come in three forms. The simplest and easiest of these is the use of grass seed that is already inoculated with arbuscular mycorrhizal fungi. The general recommended application rate is 1 kg of seed per 25 $m^2$. The quantity of inoculum present will have been adjusted to ensure a rapid and efficient colonization of the developing plants. A granular product also exists and this can be distributed with a spreader at the same time as the grass seed is sown. This type of application requires treating the area being seeded at least twice, once with the inoculum and once with a fertilizer. Because the salinity of fertilizers can damage mycorrhizal fungal propagules, it is important not to mix the fertilizer and the inoculum together in the same container. This type of inoculation is best suited to smaller areas.

The third form of inoculum is better suited for larger areas, and takes the form of a wet-able powder formulated for use by professionals. This powder is applied as a suspension, but the liquid must be continually shaken during application to avoid sedimentation in the bottom of the container.

Commercial turf producers have found that the cut and rolled strips sold for establishing lawns are more resistant to tearing when the grass is well colonized by mycorrhizal fungi. The kilometres of hyphae present in such strips link the roots together and are resistant to traction, which undoubtedly helps reinforce the turf.

CHAPTER 10

# Mycorrhizas in agriculture

As highlighted earlier, molecular biology and evidence from fossil records allow us to situate the appearance of arbuscular mycorrhizal fungi at approximately 450 million years ago, which is concurrent with the appearance of land plants (see Chapter 2). In certain fossil plants, such as *Rhynia* and *Asteroxylon*, there are remains of a fungus with a morphology similar to that of present day arbuscular mycorrhizal fungi. The subsequent co-evolution of plants and mycorrhizal fungi had enormous consequences and led to the successive divergence and speciation of plants and fungi, which allowed the development of the different types of mycorrhizas that we know today. In parallel to this, soils also evolved. In particular, the gradual weathering of the mineral parent material led to the release of soluble mineral nutrients necessary for plant growth. The undeniable role of mycorrhizal fungi in the nutrition of their hosts, allows us to understand how they were essential for the sustainable colonization of dry land by higher plants. When man appeared on earth several million years ago, plants and symbiotic fungi already had a long history of life together. For a long time, man remained a hunter-gather, and it was only about 12,000 years ago, in the Neolithic, that *Homo sapiens sapiens* first started to develop a rudimentary form of agriculture. As the vast majority of plants rely on mycorrhizal symbioses, the wild plants that become crop plants would have formed arbuscular mycorrhizal associations with fungi in the Glomeromycetes.

To fully understand the potential of mycorrhizas in agriculture, it is essential to comprehend the main features of this symbiosis and those of modern agricultural techniques, in both developed and developing countries.

## The fundamental concept of the symbiosis: the plant-fungus-soil trilogy

Being obligate symbionts, arbuscular mycorrhizal fungi need to colonize host plants to develop, and to complete their life cycle. During the establishment of the symbiosis, the fungal mycelium colonizes a host root, developing exchange organs (i.e., arbuscules) and storage organs (i.e., vesicles) in the root tissue. Furthermore, it develops an extraradical mycelium in the adjacent soil, on which spores are formed. The vesicles, mycelium and spores allow the survival of the fungus in the absence of a plant host and assure the temporal continuity of the symbiosis. These structures, which can all lead to root colonization, are referred to as propagules. Together they make up the *mycorrhizal inoculum potential of soil*. Due to both climatic and cultural factors, this potential fluctuates during the season, and from year to year. Grasping the notions behind the concept of the mycorrhizal inoculum potential of soil (see box) is essential if the effects of man on agroecosystems, and on mycorrhizal fungi, are to be fully understood. In addition, it is also important to be aware of this concept if the conditions enabling the use of mycorrhizal fungi in agriculture are to be successfully integrated into modern cropping systems.

During the course of evolution, plants have developed diverse morphologies and organs. Useful traits were naturally selected for, allowing plants to become better adapted to their environment, and to meet their needs regarding mineral nutrient uptake. It is obvious that the needs of a meadow grass are not the same as those of a tree. The extraradical mycelial network that mycorrhizal fungi develop around their host plant roots is an essential link between the plant and the soil, supplying mycorrhizal plants with, among other things, mineral nutrients. The role of this network is even more important for plants such as asparagus, which have coarse, poorly branched roots with few root hairs (magnolioid roots). In contrast, grasses, which have fine heavily branched roots (graminoid roots), are better adapted to explore small pores and cracks in the soil for nutrients and water. Studies have shown that root morphology is closely associated with the response of the plant to inoculation with mycorrhizal fungi. This response is known as mycorrhizal dependency, and is a genotypic characteristic that varies at both the plant species and cultivar level. The *mycorrhizal dependency of plants* is a working concept that allows the role of mycorrhizal fungi to be considered in low-input sustainable cropping systems. Certain plants, such as the legumes (e.g., pea, bean, vetch, soya, alfalfa and clover), have a

## THE MYCORRHIZAL INOCULUM POTENTIAL OF SOIL

The mycorrhizal fungal population of a natural soil is generally abundant and species-rich. Plants growing in such soil rapidly form symbiotic associations with one or more species of mycorrhizal fungi. However, diverse cultural techniques modify mycorrhizal fungal diversity and reduce the quantity of propagules present in the soil. This then increases the time needed for root colonization to occur, and under many cropping systems mycorrhizas are not established until the end of the plant growth cycle, which is of no benefit to the plant, or the farmer. Under certain intensive monoculture systems, where the use of phosphate fertilizer is high, the number of propagules may decrease to such an extent that mycorrhizal fungi disappear completely.

Under current agricultural practices, a range of soil parameters are regularly assessed, these include pH, mineral nutrient content, and the amount of soil organic matter. Given the role of mycorrhizas, and the conscious move towards more sustainable agriculture, it will also be important to start determining the mycorrhizal inoculum potential of soils. The technique that is most often used to determine this parameter is soil dilution. To do this, highly mycorrhizal dependent plants are grown in samples of field soil mixed with different amounts of sterilized field soil. As the soil is diluted, the quantity of fresh soil diminishes and with it the number of viable mycorrhizal propagules. After a number of weeks, it is possible to determine the number of mycorrhizal seedlings using a microscope, and to estimate the number of active fungal propagules per unit weight of soil using a probability table. Although little statistical evidence exists, studies suggest that one arbuscular mycorrhizal fungal propagule per gram of soil is an acceptable level for agricultural purposes. This knowledge is fundamentally important, and the evaluation of natural inoculum levels should be encouraged.

This useful technique has, up until now, found few uses outside the laboratory. In the future, advances in molecular biology are likely to provide effective, accurate and lower costs tools for the evaluation of the mycorrhizal inoculum potential. However, for the moment we will have to make do with the only technology that exists.

high mycorrhizal dependency, while others have evolved without the need for it (e.g., beetroot, canola, cabbage and turnip). A little later, we will deal with the concept of the *mycorrhizal receptivity of soil.*

Soil provides plants with a support, water, and mineral nutrients. In natural, undisturbed, closed ecosystems, mineral nutrients are constantly recycled and the quantity of biomass produced is proportional to the availability of mineral nutrients released by the weathering of rock. When man tries to grow crops in such systems, he often uses plants that are poorly adapted to low soil fertility. Furthermore, because the cultivated plants require more nutrients than the soil can sustainably provide, there is a subsequent degeneration of the environment and a loss of soil fertility. Cropping results in the opening of closed systems, and commercial exchange contributes to the exportation of mineral nutrients. We know that mycorrhizas enhance the mineral nutrition of plants, particularly with regards to phosphorus, which, together with nitrogen and potassium, is one of the three major elements needed for plant growth. Interestingly, the level of phosphorus in rock is very low, being 16 times less abundant than potassium. Over the centuries, mineral nutrients have been extracted from the countryside and concentrated around villages and towns, and for a long time forests have acted as a mineral nutrient reservoir. Up until the 17$^{th}$ century, mycorrhizal fungi probably played a fundamental role in crop production, because phosphorus was the limiting factor for plant growth. Productivity under early agriculture was very low – for example, in the 10$^{th}$ century, the rye harvest was 1:1 (i.e., just enough to re-sow) and the wheat harvest was between 200 and 300 kg per hectare, even though it had reached 1800 kg per hectare 4400 years ago. This period in history was marked by a series of important famines and the collapse of highly evolved civilizations.

Table 10.1 shows the detailed results of a field experiment investigating the growth of several plants in a soil containing a natural mycorrhizal flora (M) and in the same soil after sterilization (NM).

The third column shows the relative field mycorrhizal dependency (RFMD, calculated as M-NM/M) of the plants tested, which varies from 0 to 100%. This shows the effect of arbuscular mycorrhizal formation for a given field site and a given plant (e.g., 99.2% for carrots, 59.2% for tomatoes, and 0% for wheat). These data were obtained at a soil phosphorus ($P_2O_5$) level of 100 ppm (parts per million).

Interestingly, at a theoretical soil phosphorus level of 50 ppm, the value of the RFMD increases from 95.7 to 100% for leek, and from 0 to 20% for wheat. If the soil phosphorus level is further reduced to 25 ppm, the RFMD of wheat could reach between 40 and 60%. In contrast, if the soil phosphorus level is increased to 150 ppm, the RFMD of carrot decreases, and becomes non-existent for potato. The RFMD is a parameter that varies as a function of the soil phosphorus content, from which come the concept of the mycorrhizal dependency of plants; the plants being more or less dependent on mycorrhizas to satisfy their phosphorus needs, relative to the quantity that the soil can provide.

**TABLE 10.1**

| THE MYCORRHIZAL DEPENDENCY OF PLANTS | | | | | |
|---|---|---|---|---|---|
| | | | RFMD % | | |
| | | | MEASURED | THEORETICAL | THEORETICAL |
| PLANT | MASS (g) | | $P_2O_5$ | $P_2O_5$ | $P_2O_5$ |
| | M | NM | 100 ppm | 50 ppm | 150 ppm |
| Carrot | 9.2 | 0.07 | 99.2 | 99.5 | 60 |
| Pea | 40.3 | 1.3 | 96.7 | 97 | 55 |
| Leek | 11.9 | 0.5 | 95.7 | 97 | 50 |
| Bean | 13.3 | 0.7 | 94.7 | 95 | 50 |
| Broad bean | 21.8 | 1.4 | 93.5 | 94 | 45 |
| Sweet corn | 166.5 | 45.5 | 72.7 | 80 | 35 |
| Tomato | 174.6 | 71.2 | 59.2 | 65 | 10 |
| Potato | 185.3 | 107.5 | 41.9 | 60 | 0 |
| Wheat | 155.5 | 155.6 | 0 | 30 | 0 |

This dependency ranges from almost total (e.g., carrot) to non-existent (e.g., wheat), depending on the species in question. It is not only linked to the phosphorus needs of the plant and the soil phosphorus availability, but also to the capacity of the plant to access available phosphorus. Those plants with coarse, weakly ramified, root systems (e.g., carrot, leek and onion) are hardly able to grow without arbuscular mycorrhizal fungi, even in soil that has a relatively high available phosphorus content. In the same soil, plants with fine, heavily branched, root systems (e.g., wheat) are able to grow without mycorrhizas, unless the soil phosphorus level is particularly low. For tests conducted in the field, it is important to refer to the RFMD, which should allow the phosphorus fertilizer application level to be adjusted to suit the species being cultivated in a given soil. This approach holds the key to reducing fertilizer applications and, therefore, a move towards more sustainable agriculture.

## The move to large-scale intensive cropping systems

Towards the end of the Middle Ages, a profound agricultural transformation took place in Europe. Up until then, the tradition farming method had been to cultivate half of the land and leave the other for grazing. On the cultivated portion, the soil was worked by hand or using a man-powered swing plough. Half of the cultivated land was used to grow cereals and the other half was left fallow. This was small-scale agriculture. The development of larger ploughs led to the adoption of the horse as a draught animal; however, the diet of a working horse needs to be supplemented with cereals (e.g., barley or oats). As a result, in the north of France and the Paris Basin, the land used for growing cereals increased from 25% under the small-scale farming system, to 33%. This was the origin of the term large-scale farming that we still use today. The swing plough was a tool that literally only scratched the soil surface, preparing a sowing trench and slightly crumbling the adjacent soil. Furthermore, it did not break up existing root systems or the extraradical mycelial networks. In contrast, larger ploughs are dissymmetric tools that turn the soil, bringing the nutrient elements to the surface and favouring root growth. This fundamental change in agriculture, introduced a technique that actively disrupted mycorrhizal fungi by destroying mycelial networks and burying spore-rich surface horizons deeper in the soil profile. From another angle, ploughing loosened the soil and favoured root development, and plants were less dependent on mycorrhizal fungi to satisfy their mineral nutrient needs. In addition to this, another factor that was detrimental to arbuscular mycorrhizal fungi was the progressive abandoning of fallow land, which had served as a reservoir of mycorrhizal fungal propagules.

## Modification of the environment by man

The passage from small- to large-scale faming did nothing to slow the decline in soil fertility. On the contrary, increased production further amplified the export of mineral nutrients and enhanced their concentration around towns. In France, this negative balance continued up until the beginning of the 20$^{th}$ Century. The first agronomists (a term originally coined in 1798 to describe certain members of the *Société d'Agriculture de Paris*) confronted the idea that plants obtained nutrient directly from organic matter, but it was only in 1840, following the publication of a book by Liebig, that the mineral nutrition of plants started to be widely accepted. With regards to phosphorus (discovered by Brandt in 1664), the first beneficial effect of this element on plant growth was observed using guano from Peru. However, the commercial importation of this substance for use as fertilizer did not start until 1920. The phosphorus exported by livestock farming is mainly due to it being stocked in bones, and these were the next potential source of phosphorus to be investigated. In 1810, Murray developed the technique of treating bones with acid to create superphosphates. In 1843, Gasparin reported that the *Société d'Agriculture de Paris* proposed to use the earth from cemeteries as a fertilizer and that skeletons were collected after the Battle of Waterloo, for this same reason. The use of mineral phosphates increased from 1890, but it was only after the Second World War that the use of chemical fertilizers became the norm.

The industrial revolution in the 19$^{th}$ century, led to a rapid expansion of intensive agriculture and this has profoundly modified the soil environment. For example, soils have been enriched with massive applications of organic matter, their pH has been modified using lime to improve certain chemical properties, they have been drained to improve their physical properties, and their nutrients levels have been increased using massive quantities of diverse chemical fertilizers.

In parallel, in an attempt to increase crop yields there has been a constant effort to improve the plants cultivated, and to enhance their resistance to disease organisms and environmental stresses. However, these new high-yielding cultivars have tended to need greater amounts of chemical fertilizers and pesticides. Furthermore, in many cases, technical and economic pressures have forced farmers towards large-scale monocultures of cereals, which have little dependence on mycorrhizal fungi. Moreover, modern cultivated plants are less dependent on mycorrhizas than their ancestors were. This situation is almost certainly due to the fact that crop plant breeders have always worked in heavily fertilized experimental stations; this was to avoid the possibility of low nutrient levels limiting experimental results. Further more, disease resistant plants have always been preferentially selected for, and particularly those resistant to fungal disease. This has almost certainly progressively led to certain plants becoming less receptive to mycorrhizal fungi.

Studies over the last 50 years have highlighted the fact that mycorrhizas have a greater impact on plant growth when the soluble phosphorus content of the soil is moderate or low (i.e., generally less that 100 ppm). This is not surprising, as they have co-evolved with plants in soils that are naturally low in available phosphorus, due to the low phosphorus content of the parent material. Phosphorus is a non-toxic element for plants, fungi and man. Plants only absorb ions in solution, and this absorption takes place, mainly through the root hairs, from a zone very close to the root surface (1 to 2 mm). However, plants with mycorrhizas also absorb ions via the extraradical mycelium, which can extend several centimetres out from the root. When a soluble phosphate fertilizer (e.g., superphosphate) is applied, it enriches not only the soil phosphorus pool but also, more importantly, the concentration of phosphate ions in the soil solution. The diffusion coefficient of phosphate ions is between 100 and 1000 times slower that potassium, and between 10,000 to 100,000 times slower than that of nitrogen. This coefficient, which depends on the ratio of the quantity of ions fixed over the quantity of ions in solution, is increased by the addition of soluble fertilizers. This, in turn, increases the overall speed of diffusion towards the roots and improves plant nutrition. As a result, plants have less need for mycorrhizas, at least with regards to phosphorus nutrition. Nevertheless, the role of mycorrhizal fungi should not be reduced to the simple question of nutrition, because they also play an important role in resistance to disease, drought and other environmental stresses.

## Operational concepts of modern agriculture: the cropping system and land management practices

The notion of cropping systems appeared at the end of the 18$^{th}$ century and the concept has evolved in the context of increasing agricultural modernization. In 1975, considering the production system to be the overall agricultural exploitation, the working group at INRA-INAP in France proposed the following definition for cropping systems: under the whole of the production system, the cropping system is defined as a surface of land treated in a homogeneous way, and includes crop rotation and land management practices.

Ploughed land, mowed or natural meadows, and orchards are all cropping systems. In highly specialised large-scale cereal production, particularly under monocultures, the cropping system and the production system are difficult to differentiate. In mixed agriculture that combines crop production and livestock rearing, several systems coexist.

Land management practices comprise the logical series of events or treatments that are applied to a species of cultivated plant, and these range from soil preparation and seeding, to harvesting.

Whether or not to use crop succession or rotation on a given soil is the farmer's decision, and is controlled by specific rules (e.g., soil preparation, the species and cultivar used, and fertilizer application). This is because at any one time, a given crop is under the influence of the previous crop grown on the site (e.g., an increase in nitrogen availability due to a legume, or an increase or decrease in weeds after a cereal or weeded crop, respectively) and will, in turn, influence the next crop to be grown.

It is easy to understand that these two concepts (i.e., the cropping system and management practices) are the key to the functioning of agriculture. Moreover, the practices used, profoundly alter the soil environment, including its microfauna and microflora populations, which include arbuscular mycorrhizal fungi.

## Mycorrhizas in cropping systems

In the field, the role of the mycorrhizal symbiosis is expressed through the mycorrhizal dependence of the plants and the mycorrhizal potential of the soil. These two concepts are reciprocally linked, as the latter is a result of the expression of the former, which also governs the latter. The development of mycorrhizas in a given agricultural field is largely dependent on the type of cropping system and, in particular, the rotation of plants used. This is because certain plants are more mycotrophic than others.

The factors that are specific to a given cropping system include modifications made to the plant (e.g., the selection of a given variety, genetic improvement, pesticide or growth hormone application, and the use of fungicide treated seed) and modifications made to the soil (e.g., fertiliser, pesticide or lime applications, drainage, irrigation, and deep or shallow ploughing, or no-tillage approaches).

### Modification to plants

There is no doubt that crop breeding programs conducted on nutrient-rich soil that sought to create new high-yielding varieties have, inevitably, selected genotypes that are less mycotrophic than the original stock. Furthermore, some may have even lost the need to form mycorrhizas. This is due to the fact that after successive selection programs, plants loose certain genes and these can include genes that are important for the establishment and functioning of mycorrhizas. Because of this, modern varieties of wheat and tomatoes now have a mycorrhizal dependency that is much lower than that of their ancestors. In addition, maize plants selected for their resistance to fungal pathogens are less receptive to mycorrhizal fungi than their more susceptible parent plants.

Furthermore, the fungicides that are regularly used to treat seeds, are much more damaging for arbuscular mycorrhizal fungi than those applied by spray after root colonization has been established. The toxicity of a fungicide to mycorrhizal fungi depends on the molecule used, the amount applied and the timing of treatments. Interestingly, certain fungicides (e.g., fosetyl-Al, metalaxyl and promocarb) have no negative impact on mycorrhizal fungi and some may even enhance root colonization by these symbiotic fungi.

Herbicides are another group of pesticides that are widely used in agriculture. These are used to destroy weeds and are generally not toxic to mycorrhizal fungi. Nevertheless, these chemicals destroy many potential host plants and so contribute to an overall reduction in the mycorrhizal inoculum potential of a soil.

## Modifications to soil

Preparation of the soil prior to seeding modifies the distribution of mycorrhizal fungi in the soil. Deep ploughing buries arbuscular mycorrhizal fungal spores deep in the soil profile, and this potentially places them out of the reach of shallow rooted crop plants. Furthermore, pulled harrows break up the mycelial network left by previous crop plants. In contrast, when disc harrows are used to break up the soil, they fragment the roots of the previous crop. If these roots were of a mycorrhizal plant, this technique can help increase the number of propagules and the mycorrhizal inoculum potential of a soil. Studies have shown that the destruction of the mycelial network of a previous crop, results in a reduction in the absorption of phosphorus by seedlings that subsequent develop in the soil. However, if the density of soil inoculum is high, this is only a short-term effect.

The practice of fertilizer application was developed to complement the mineral nutrient pool of a given soil, bringing it up to a sufficient level to satisfy crop plant needs. Although this technique has been used to excess in many parts of the world, in certain countries, soil fertilization is essential if the food needs of the population are to be met. In the case of phosphorus, which is strongly and rapidly fixed onto components of the soil matrix, the approach that was favoured for many decades was the progressive saturation of exchange sites. This was done on the understanding that once these were saturated, any additional applications would be directly available for plant growth. However, scientific research has highlighted the fact that this approach can have environmentally catastrophic consequences and is economically unrealistic. In addition, the formation of mycorrhizas, and as a result the abundance of mycorrhizal fungi, is tightly linked to the level of available phosphorus in the soil nutrient pool and, in particular, to the level of phosphate ions in the soil solution. The colonization of root systems by mycorrhizal fungi, the mycorrhizal inoculum potential of soil, and the mycorrhizal dependency of plants, are all inversely correlated with the concentration of phosphate ions in the soil solution. Furthermore, this result is not linked to a given form of phosphate-rich fertilizer, or whether the fertilizer is organic or mineral, because plants only absorb ions in solution. Moreover, higher concentrations of mineral nutrients in the soil solution result in well-fed plants, which typically has a negative affect on the level of mycorrhizal fungal colonization in their roots. In certain situations, the level of phosphorus is so high that it becomes incompatible with the formation of mycorrhizas.

Although pesticides, fungicides and soil disinfectants are often unfavourable to mycorrhizal fungi, the negative effect of high levels of available phosphorus is direct and often results in a total elimination of these fungi from a given site.

## Managing mycorrhizal fungi in the field

There are two approaches to managing mycorrhizal fungi in cultivated soils: a technological approach that aims at introducing selected strains of naturally-occurring fungi, and an agroecological approach that aims to manage the existing indigenous fungal populations. In both cases, the fungi are affected by the cropping system used.

### Technological approach: selected strains

Numerous studies have clearly shown the advantages of adding mycorrhizal fungi to a wide range of agricultural soils, and that these fungi have no negative impact on beneficial soil microbial populations. The introduced micro-organism must find a suitable ecological niche, where it can grow and compete successfully with others for mineral nutrients. The introduction of a biological aid for crop plants is ecologically successful if it survives, but it is only successful from an agricultural point of view if it develops enough to produce the effects that were hoped for.

The commercial production of arbuscular mycorrhizal fungal inocula was discussed in Chapter 8. The development and production of effective inocula has taken a considerable amount of time, effort and resources. The first high-quality inocula were expensive and difficult to obtain, in particular in developing countries, where they were perhaps most needed. Nevertheless, in recent years, a number of new breakthroughs have been made, and products obtained under sterile *in vitro* conditions have become widely available in Canada and India. In Canada, several tens of thousands of hectares under wheat, lentils and soya were inoculated with commercially produced arbuscular mycorrhizal fungal inocula in 2008. While on the other side of the globe, over 10,000 farmers in India buy mycorrhizal inocula every year, for use in small-, medium- or large-scale cropping systems. Currently, there are several companies throughout the world that are working to produce and formulate inocula for large-scale agricultural applications. In Cuba, non-sterile products have been developed, and their use has spread to South America. Recent technological advances have allowed the development of large-scale *in vitro* inocumlum production in Canada and India (visit the official Internet sites of Teri Industries and Premier Tech for further details, see page 131 for the addresses). The production cost has been substantially reduced, making mycorrhizal inoculation cost-effective for both horticulture and large-scale crop production systems even in developed countries.

## Agroecological approach: indigenous strains

The indigenous mycorrhizal fungal strains of a given agroecological system are often as efficient as some of the commercially available strains. However, even when they are not quite as efficient, they are at least adapted to site-specific environmental conditions. The agroecological approach is less costly than the large-scale inoculations used under the technological approach, but can be complicated. Careful management must be done to ensure that the mycorrhizal inoculum potential of a given soil is successfully increased and stabilized. This can be achieved through the use of appropriate agricultural practices. This implies a strict control of the amount of phosphorus fertilizer and pesticides applied, the employment of crop rotations that exclude or minimize the use of non-mycotrophic plants, a reduction in soil disturbance, and the use of cover crops and intercropping. Under such systems, the notion of mycorrhizal dependency of plants is all-important. It should be noted that the above does not exclude the possibility of investigating the efficiency of the indigenous fungal strains.

The level of plant available phosphate in a given soil remains the most important factor for the management of populations of mycorrhizal fungi. If it is compatible with the formation of mycorrhizas, then it is the concept of crop rotation using mycotrophic plants that must be mastered. Indeed, for a soil with the same mycorrhizal inoculum potential, colonization will be greater in a legume crop (e.g., soya, pea, bean and clover) than in a cereal crop (e.g., wheat, oat, barley and maize), and nonexistent in a crop of a member of the non-mycorrhizal Chenopodiaceae (e.g., beetroot and spinach) or Cruciferae (e.g., cabbage, canola, mustard and turnip). When considering crop rotation, the preceding crop will greatly influence the mycorrhizal status of the crop to be planted. The yield of maize has been shown to be a function of the cropping history of the field and, in particular, it is directly correlated with the percentage of mycorrhizal colonization of the previous crop. Furthermore, long-term trials have shown the negative effect of a build-up of phosphorus (of mineral or organic origin) and crop rotations using slightly or non-mycotrophic plants, on the mycorrhizal inoculum potential of soils.

Under modern cropping systems in developed countries, the technique of allowing land to lie fallow has been largely abandoned. However, long-term trials have shown that this approach is beneficial for the management of mycorrhizal fungi. In contrast, the use of fallow is still an essential component of the cropping systems in many African countries. Nevertheless, current demographic pressures are forcing the length of the fallow period to be reduced from ten years, to only two or three. Interestingly, and contrary to what might at first be believed, under pasture, the mycorrhizal inoculum potential regenerates during the first year and then decreases. This is because under pasture, deep-rooted, shrubbier, perennial species participate little in maintaining the number of propagules of mycorrhizal fungi in the surface layers of the soil, where seeds are sown.

Intercrops are often used to capture nitrates or reduce soil erosion. These are planted after the principal crop, and offer a relay system for mycorrhizal fungi in much the same way that certain weeds do. This allows the mycorrhizal inoculum of the soil to be maintained or enhanced prior to the sowing of the next crop.

## The different cropping systems

There are a number of unforeseen factors linked to soil type, climate, the economy and local culture, which govern different agricultural practices under particular cropping systems. These practices, such as soil preparation, the choice of a particular crop or variety, and soil fertilization, interact with each other. Therefore, they must all be considered in order to ensure that the efficiency of the cropping system is maintained. Indeed, altering one component leads to repercussions that might, or might not, be important. For example, growing a legume crop on a sandy soil can have an immediate and important effect on the level of nitrates in the water table. Furthermore, a single action can modify several environmental variables. For example, soil preparation alters soil porosity and structure, the location of propagules of pathogens and of symbiotic mycorrhizal fungi, and weed plant populations. The inverse is also true: a single environmental variable can be modified by several different agricultural practices. For example, the intensity of a given wheat disease may be a function of the crop rotation system, the date of sowing and the cultivar used.

Therefore, a cropping system is a coherent and complex group of factors. There are hundreds of different systems used throughout the world and these range from manual to mechanised, from simple to complex, and from subsistence to commercial. Given the consistent nature of cropping systems, which makes change difficult, what is the future for mycorrhizas?

Each cropping system is different, and may be for horticulture, cereal monoculture, mixed arable, mixed arable and livestock, or intensive dairy or beef farming. Furthermore, a given cropping system may be used at a range of altitudes, and irrigation, slash-and-burn, or fallow may, or may not, be used. Therefore, each must be considered individually when determining the possible use for mycorrhizal fungi.

Whether inoculation is being used to enhance the quantity or quality of mycorrhizal fungi present at a given site, or whether natural mycorrhizal fungal populations are being managed, the general techniques are the same as those used under the cropping system. However, these would alter slightly in cases where an annual inoculation is economically possible and agriculturally efficient.

## The use of mycorrhizas in cropping systems

For determining the conditions necessary for the use of mycorrhizas, it is important to consider the three-way interaction between the soil, the plants and mycorrhizal fungi. Before going any further, it is necessary to highlight the fact that nothing can be done by land management practices alone.

### The soil receptivity

This is a crucial criterion in determining the suitability of using mycorrhizal fungi. For example, in soils in regions where battery farming (e.g., of chickens, rabbits and pigs) or intensive dairy farming is practiced, muck spreading has led to phosphorus level that may exceed 1000 ppm. These over fertilized soils inhibit the establishment of mycorrhizas. In fact, even if additional applications were stopped, these soils would have to be cropped for over 100 years before level of phosphorus in the soil solution would be compatible with mycorrhizal formation. In regions where cropping systems are based on large-scale intensive cereal monocultures, the level of phosphorus in the soil solution is not as high. However, the present management techniques that are used seek to maintain the level of soil fertility that has been acquired. Once again, this does not favour the formation of mycorrhizas. Nevertheless, because of the low concentration of phosphorus in rock, most soils tend to be deficient in phosphorus. Furthermore, in developing countries, the use of chemical fertilizers is often not economically possible. These phosphorus-poor soils are therefore an example of a situation where mycorrhizal fungi can play a useful role.

## The plants

If the right conditions of soil fertility are met, the second factor that controls the development and the expression of the mycorrhizal symbiosis is the plant. Monocultures tend to be unfavourable for mycorrhizal fungi. This is because they are usually comprised of cereals that have low mycorrhizal dependency, and this type of agriculture often requires the massive use of pesticides. However, it is important to remember that plants such as wheat, which are only weakly dependent on mycorrhizal fungi in soils with reasonable levels of available phosphorus (e.g., 100 ppm), becomes more dependent if the level of soil phosphorus falls below 50 ppm. In certain soils, such as those developed under tropical conditions, such levels of available phosphorus are only rarely reached. From another point of view, crop rotation can be favourable for mycorrhizal fungi. This is because crops of highly mycorrhizal dependent plants, such as the legumes, can compensate for the lower mycorrhizal inoculum potential of a soil following a crop of a member of the non-mycorrhizal Chenopodiaceae or Cruciferae.

## The fungus

The entire land management protocol must be designed to protect mycorrhizal fungi; however, this can be difficult to achieve. Indeed, a single management practice, such as the application of a fungicide, can ruin attempts to rebuild a natural mycorrhizal fungal population, or cancel out the benefits of an inoculation.

It is often considered better to inoculate with local strains of mycorrhizal fungi; however, a number of studies done in Cuba, India and several countries in Central and South America have shown that introduced strains can be highly effective. For example, strains isolated in Cuba are now successfully used in a great diversity of soils, not only in the West Indies, but also in Central and South America, and strains produced under *in vitro* conditions in Canada have proved effective in Mexico.

# Sustainable agriculture based on a scientific understanding of the role of mycorrhizas

## A. Questioning current international agricultural practices

Environmental groups must be given the credit for having rapidly understood and raised awareness about some of the inherent problems associated with current agricultural practices. However these alarmist positions were often followed by irrational statements weakening their credibility.

Recently, the first meeting of the International Society of Organic Agricultural Research (ISOFAR), held in Australia in 2005, entitled *Researching sustainable systems*, and that of the International Federation of Organic Agriculture Movements (IFOAM), held in Italy in 2008, entitled *Cultivating the future based on science*, gave a great deal of importance to the integration of mycorrhizal fungi into new sustainable agricultural concepts and practices.

The rational questioning of current agricultural practices can be approached from a range of perspectives, including the energetic and environmental costs incurred, the use of fertilizers and pesticides, the dietary and neutraceutical quality of agricultural produce, and the development and maintenance of soil fertility. Moreover, as the ISOFAR and IFOAM conferences suggested, agricultural practices can also be questioned from a mycorrhizal perspective, taking into consideration the advantages that the arbuscular mycorrhizal symbiosis contributes now, and is likely to contribute in the future. The rest of this chapter briefly examines this, and suggests some of the topics that should be investigated to help integrate mycorrhizal fungi into agriculture practices in the 21st century. It should be noted that, to date, the vast majority of the agricultural concepts and practices that have been developed largely ignored the existence of mycorrhizal fungi. However, these fungi were instrumental in the colonization of dry land by plants, and they continue to play an essential role, governing virtually every aspect of the life of wild and, where permitted, cultivated plants. The time has come to take mycorrhizal fungi into consideration and to re-examine agriculture practices.

## B. The first step to rethinking agriculture in the light of what we know about mycorrhizas: host plant genetics and fertilizers

The most powerful catalyst of the first green revolution was the genetic improvement of agricultural and horticultural plants. Nevertheless, the genetic selection of new cultivars was done with an unawareness of the importance, or even existence, of mycorrhizal fungi. For example, the selection of cultivars was done under conditions of optimal or excessive soil fertility, and fungal disease resistant plants were favoured without considering the secondary effects that this might have on root colonization by mycorrhizal fungi.

In the past, the application of fertilizers, notably those rich in phosphorus, was done without considering mycorrhizal fungi. This led to the widespread adoption of excessively high fertilizer application rates, which negatively affect the functioning of mycorrhizas. The recommended fertilizer application rates, which are usually provided by the companies that produce and market them, should ideally include different application rates for use with or without mycorrhizal fungal inoculation. However, this is not current practice.

Ironically, the registration of mycorrhizal fungal inocula is subject to a series of strict tests and careful monitoring, whereas the registration of natural or synthetic pesticides does not require that their harmlessness to mycorrhizal fungi be demonstrated.

Furthermore, unfortunately, the irrefutable role of mycorrhizal fungi in the drought resistance of host plants, and the importance of mycorrhizal fungi in the development and long-term maintenance of soil fertility, continue to be largely ignored. Finally, it should be remembered that the formation of mycorrhizas profoundly modifies the biochemical composition of the host plant, which positively influences the dietary and nutraceutical quality of agricultural and horticultural plants, and their interactions with various pest species.

## C. Increasing our knowledge of mycorrhizal fungi and integrating them into global agricultural practices

Bearing in mind the limited consideration given to mycorrhizas in agriculture in the past, and the present level of our knowledge concerning the arbuscular mycorrhizal symbiosis, it is important to continually highlight the role of this association. This is all the more pertinent, given the fact that mycorrhizal fungi are likely to play an instrumental part in the next green revolution. It will also be important to increase the resources invested in improving our understanding of the functioning of this symbiosis, and to integrate current knowledge and novel discoveries into global agricultural practices.

As we have seen, the biology of mycorrhizal fungi is highly complex and quite different from that of any other group of living organisms. Certain aspects of these fungi, such as the dynamic nature of their cytology, their high level of intraspecific genetic diversity, their ubiquitous nature, their complex classification and phylogeny, their multiple interactions with their plant hosts, their synecologic interactions with other soil micro-organisms, and their role in the development and maintenance of soil fertility, make the detailed study of these organisms a huge planetary challenge.

The time has come to change the paradigms underlying the genetics-based improvement programs of the majority of agriculturally and horticulturally important plants.

With regards to arbuscular mycorrhizal fungi, given the high level of intraspecific variation, there is a huge potential for the genetic selection of isolates. This has hardly been touched upon, but recent studies have highlighted the possibility for selection and improvement of strains of mycorrhizal fungi adapted to different environmental conditions.

The obligatory nature of this symbiosis is an important challenge to companies producing and formulating inocula. However, technological advances now permit mass production of the propagules of these fungi, allowing the inoculation of hundreds and even thousands of hectares.

The challenge is just as great for those using mycorrhizal fungi, as the behaviour of individual cultivars of a given host plant species must be taken into consideration, together with the characteristics and constraints of different soils and climates found throughout the world.

If the first green revolution that took place during the last few decades of the 20th century occurred under the aegis of a certain level of genetic improvement (which today is greatly surpassed) and of the ability to fix nitrogen, the next green revolution must be founded on a solid understanding of mycorrhizal fungi and their general use in agriculture and horticulture.

It is high time that organizations such as the FAO, and governments of both developed and developing countries, understand what is at stake and integrate the concept of mycorrhizal fungi into their policies underlying the support of research and development, and the development of modern agricultural practices that are both environmentally friendly and sustainable.

## Evolution of the cropping system towards sustainable agriculture

It is now realized that the cropping systems employed in a large number of developed countries are as suitable for the use of mycorrhizas as those in developing countries. However, not all agricultural practices in low-input cropping systems in developing countries favour mycorrhizal fungi. Indeed, the technique of slash-and-burn practiced in some countries to supply mineral nutrients, and in particular the burning of plant material in trenches prior to planting, sterilizes the soil and eliminates mycorrhizal fungi.

As a result of certain reoccurring environmental problems, such as soil, water and air pollution, eutrophication, and increased outbreaks of certain animal diseases, the use of highly intensive cropping systems is being questioned. Growing public awareness of environmental issues is leading to new constraints on farmers. Increasingly, the public are seeking both a healthier environment and a greater diversity of organically produced goods. The concept of sustainable agriculture first emerged in the 1980s. Briefly, this approach seeks economically viable agricultural practices that allow agriculture to evolve in harmony with the environment, providing high-quality produce but also ensuring the long-term quality of soil and water resources. To obtain these objectives, sustainable agriculture favours an increase in the biodiversity of the crop plants cultivated, a reduction in the use of chemicals, the use of less soluble fertilizers, the recycling of organic waste, and the use of biological controls against pests. It remains to be seen whether, in this context, the evolution of cropping systems will take mycorrhizal fungi into consideration.

### The case of developed countries

New constraints concerning the use of chemicals are being imposed on farmers. There is a general consensus that the quantity of chemicals being used has exceeded an acceptable limit, but how can production levels be maintained if inputs are to be reduced? Change will be slow, especially considering that many of the heavily fertilized phosphorus-rich soils may take hundreds of years before they become suitable for the use of mycorrhizas. This is true for a large proportion of agricultural soils in Europe. Nevertheless, certain cropping systems, such as those in upland areas, have not attained the same level of intensification as the large-scale cropping systems. Furthermore, in certain Europe countries, horses are still used to pull farm machinery and, generally speaking, the production of grass for grazing and hay is done with little or no chemical input. Several studies have shown the possibility of controlling the floristic composition of such environments by using their mycorrhizal dependency. In Europe, certain regions of low-intensity livestock farming, whether for meat or dairy products, have a labelling system that guarantees the origin and quality of the product. Strict norms are used to supervise agricultural production in these regions and the agricultural practices employed are compatible with the use of mycorrhizal fungi. Indeed, the plants present must be mycorrhizal, but few studies have been done.

Nevertheless, not all developed countries have reached the same level of intensification in terms of the cropping systems they use, this is often simply due to the availability of extensive areas of cultivatable land relative to the country's population size. For example, in the Americas, the soil of the vast Canadian prairies has never received important quantities of fertilizers or pesticides. In such situations, which also occur to different degrees in other countries, there are opportunities to use mycorrhizas, particularly in cropping systems that include a legume in their rotation.

Throughout the world, there are vast areas of farmland where a green revolution based on environmental respect and inoculation with mycorrhizal fungi would allow more cost-effective, sustainable and environmentally friendly agricultural practices to be envisaged.

In France, an important reduction in the sale of phosphate fertilizers has occurred over the last 15 years; passing from 1,500,000 tonnes in 1990 to 700,000 tonnes in 2005. However, this is partly due to the fact that soils under intensive livestock farming are constantly enriched with organic matter, which contains mineral elements from imported food supplements (e.g., oil cake). Nevertheless, this reduction is an important first step, and in zones of less intensive agriculture, or where agriculture is undergoing de-intensification, there are soils where mycorrhizal fungi could be used.

When considering the potential use of mycorrhizal fungi, it is also important to keep in mind the fact that the world's phosphorus reserves are rapidly being exhausted. Furthermore, the financial and environmental costs involved in transporting fertilizers from one continent to another are becoming increasingly hard to justify.

Along similar lines, the social acceptability of chemical pesticides is being questioned throughout the world. Europe is the largest consumer of pesticides, using 28% of the global production, closely followed by North America and Asia, which use 25%, and Latin America, which uses 18%. This is in sharp contrast to

Africa, which only uses 4%. The most frequently used pesticides are herbicides, which represent about 50% of all the pesticides produced. Although these are the least toxic pesticides for man and animals, the amounts applied are so large that they frequently cause important reductions in water quality. For this reason, there are one of the main targets in the effort of reduce the amount of chemicals used in agriculture. Bearing this in mind, cropping systems will have to be rethought and restructured. The reliance on chemicals to combat weeds will have to be reduced, and the use of mechanical (e.g., soil preparation) and biological control (e.g., helper plants and organisms, and allelopathy) methods will necessarily increase. The latter two methods cannot completely eliminate weeds, but they can aim to maintain weed populations below a limit that negatively impacts crop production. If this change in land management does occur, it will be extremely useful for maintaining the mycorrhizal inoculum potential of soil. Indeed, if a certain density of weeds is to be tolerated during the production of a given crop, these plants will be able to serve as a relay for mycorrhizal fungi. The presence of weeds will be particularly important if a non-mycotrophic crops is grown, or if intercropping is not used.

## The case of developing countries

Although the problems associated with using arbuscular mycorrhizal fungi in agriculture in developing countries are radically different, some practices can also lead to the virtual disappearance of mycorrhizal fungi. While the intensification of food-producing crops has not occurred, the establishment of large-scale plantations for industrial crops (e.g., cotton and rubber) or fruit production (e.g., banana and pineapple) rely on huge quantities of chemicals. These plantations, which are frequently owned by foreign investors, are examples of intensive agriculture, and the conditions for using mycorrhizal fungi are subjected to the same constraints as in developed countries.

In many countries, subsistence agriculture for food and income is still completely manual. Furthermore, the use of chemical inputs, in particular mineral fertilizers, is limited or inexistent. Which means that the soils, especially tropical ones that are naturally poor in organic matter and available mineral nutrients, remain nutrient poor. Moreover, crop harvesting further depletes the soil nutrient pool. Populations in developing countries continue to grow and there is an increasing demand for food. This is forcing an increase in the surface area under agriculture, a reduction in the length of the fallow period, and a reduction in the amount of land available for grazing. This has a knock on effect that results in a reduction in herd size, a reduction in the amount of organic fertilizer, and a subsequent reduction in crop yields. In countries such as Africa and India, the soils are often too poor to meet the food needs of the population. Furthermore, often all the cultivatable soils are already farmed and in some cases the area under cultivation is being constantly reduced due to erosion and increased salinity. Low nitrogen availability in these soils limits yield, and adding nitrogen fertilizers can increase crop production. However, African countries in particular lack sufficient quantities of organic fertilizer. Nevertheless, the problem of nitrogen can be solved, at least in part, by low-level applications of mineral fertilizers and the use of nitrogen-fixing legumes. All tropical soils are largely lacking in phosphorus, and this becomes the limiting factor for agricultural production. However, certain developing countries have natural phosphates reserves and, indeed, are producers and exporters of this resource. It is in these countries that the use of mycorrhizas would be most judicious and efficient. In fact, mycorrhizas need a certain amount of phosphorus to work. By coupling mycorrhizas and natural phosphates, it should be possible to improve crop production in these countries. Nevertheless, it must be highlighted that even if studies under controlled conditions are encouraging, there is still a need for field trials that demonstrate the advantage of such techniques to farmers. Once this has been achieved, the economic problem associated with the guaranteed availability of free natural phosphates and mycorrhizal inoculum, still has to be overcome. In an ideal world, the latter could be distributed by non-governmental organizations.

Certain cultural practices, habits, and customs in developing countries do not necessarily help in the protection of the environment. For example, the mycorrhizal inoculum potential of certain cultivated soils in Madagascar is zero, even though they have never received any chemical inputs. There are two possible reasons for this. First, the soil is so poor that the plants are nutrient deficient and unable to favour the development of functioning mycorrhizas. Under such conditions arbuscular mycorrhizal fungi can act almost as parasites. Second, the common practice of slash-and-burn sterilizes the soil to a certain depth.

Tropical soils are for the most part fragile, and agricultural techniques that are commonly used in northern countries (e.g., ploughing) have often had catastrophic effects in these agroecosystems. In tropical soils it is important to favour direct sowing techniques, and to keep soil manipulation to a strict minimum. Where possible, sowing should be done under a cover of either dead or living vegetation. Dead vegetation, such as dry grasses, provides a mulch that mineralizes slowly on contact with the soil, supplying a small amounts nitrogen. Furthermore, it also controls the development of weeds, the loss of water by evaporation, and soil erosion. In the case of a living vegetation cover, the plants are frequently nitrogen-fixing legumes (e.g., *Arachis pintoi* and *Crotalaria* sp.). These plants are low growing and do not compete with the principal culture (e.g., paddy rice or maize). These and other techniques that limit soil disturbance, and which use legumes of high mycorrhizal dependency, are very much suited to the development of mycorrhizas and of the mycorrhizal inoculum potential of the soil.

Developing countries are likely to benefit greatly from the use of mycorrhizas. Nevertheless, even in these countries changes will need to be made to cropping systems, habits and customs, the training given to farmers, and the implication of governments and non-governmental organizations.

In contrast, it will be difficult, if not impossible, to use mycorrhizal fungi in cropping systems that rely heavily on massive inputs of chemicals (i.e., fertilizers and pesticides), that are on soils that have high levels of plant-available phosphorus, and that regularly include non-mycotrophic plants in crop rotation systems. However, novel cropping systems that use fewer chemicals and that rely to a greater extent on helper organisms, are being investigated for use in both developed and developing countries. The use of mycorrhizas is highly appropriate in such systems.

CHAPTER 11

# Production of ectomycorrhizal inocula

Forests dominated by ectomycorrhizal trees typically produce a huge number of fruitbodies of a wide range of ectomycorrhizal fungal species, and these produce an astronomical amount of spores. In contrast to arbuscular mycorrhizal fungi, these spores are small and often wind-blown. This gives rise to a spore rain that falls further downwind, even several kilometres from the source. Unless forest tree nurseries in such areas use soil fumigation techniques, the seedlings they produce automatically develop mycorrhizas. These are typically formed by a suite of species known as early stage ectomycorrhizal fungi, which includes members of the genera *Laccaria* and *Hebeloma*. In such situations, inoculation with selected ectomycorrhizal fungi is impracticable.

However, there are numerous situations where inoculation is necessary and these will be further discussed in Chapters 12 and 13, which treat the use of ectomycorrhizas in forestry and in soil restoration, respectively. Although ectomycorrhizal roots can be used as an inoculum, ectomycorrhizal fungal inoculum is typically either spore- or mycelial-based.

## Spores as a source of inoculum

The use of spore-based inocula is the simplest, most economic, and most efficient method for inoculating tree seedlings. This method requires a little planning, a sufficient quantity of spores of the species to be used, and a minimum amount of equipment.

The most convenient species to use for this are members the Gastromycetes, and particularly those within the genera *Pisolithus* and *Rhizopogon*. The fruitbodies of these species, which are more or less round, produce huge quantities of spores inside a tough envelope that only opens when the spores are fully mature.

Once the fruitbodies have been located, collecting the spores is a relatively simple process. Briefly, the fruitbodies must be regularly monitored and then harvested just before they reach maturity and open. The fruitbodies should be cleaned to remove adhering soil, dried, and opened to release the spores. The spores should then be sieved to remove pieces of the outer envelope of the fruitbody that could interfere with the inoculation process, and stored in a cool (4 °C) dry place prior to use.

If inoculation is being done in a traditional production system that does not use a modern irrigation or fertilization system, crushed dried pieces of fruitbody can be simply mixed into the growing medium at the rate of a few grams per square metre.

Generally speaking, inoculation with spores works best in plant production systems producing container-grown plants, under shelter, and with controlled fertigation. Under such systems, a suspension of sieved spores can be applied using the fertigation system, again at the rate of a few grams per square metre. Sieving of the spores is essential to avoid blocking the nozzles of the fertigation system. To ensure good colonization rates, the fertigation regime must be compatible with the establishment of the symbiosis, and excesses of nitrogen and phosphorus must be avoided.

In the case of truffles, plants may be inoculated by dipping their roots in a clay-based slurry containing mature spores.

## Mycelial inocula

Although the use of spore-base inocula is the simplest and most efficient, it is not suitable for all situations. For example, it may be necessary to use a particular strain of a given species, or a species that typically only produces a few small fruitbodies and thus a limited quantity of spores.

### *In vitro* culturing

Two principal methods are used for obtaining strains *in vitro*: the use of flesh from a single fruitbody, or the use of spores.

The *in vitro* culture of fungal mycelium closely resembles the *in vitro* culture of plants, except that the artificial medium must contain sugar. The culture medium containing the necessary sugars and mineral nutrients are solidified with a gelling agent, typically agar. The isolation technique is simple, but must be done under sterile conditions. The surface of the fruitbody should first be cleaned and then a shallow (1 to 2 mm) incision made all the way round it. The fruitbody should then be broken in half to expose the inner flesh.

Still working under sterile conditions, a small piece of flesh, ideally not more than a few square millimetres, should be transferred to a Petri dish containing the chosen solidified medium. The Petri dish is then incubated in the dark under controlled conditions. After several days or several weeks, depending on the species, the mycelium grows out of the transferred tissue and onto the medium, where it generally forms a circular colony. This culture can then be multiplied by transferring a small amount, again under sterile conditions, to another Petri containing fresh medium.

Obtaining a sterile culture from spores is a little more complicated. Ideally, spores should first be obtained in a sterile Petri dish. For gilled species, this can be done by fixing a small piece of the fruitbody cap (with the gills pointing downwards) onto the inside of the lid of the dish. Once the spores have been shed, these can be suspended in sterile water and a small drop can be used to inoculate a Petri dish containing the same solidified medium as used above. The Petri dishes should be incubated as above and observed daily under a microscope. Germinating spores should be rapidly transferred to fresh medium to avoid contamination by mycelium from opportunistic fungi that were present on the piece of fruitbody used to produce the spores.

The use of this method, results in monosporal cultures, and each is of one of the two fungal mating types (monokaryons). To obtain a suitable strain for mycorrhizal work, monosporal cultures must be crossed to obtain dikaryons. By controlling the monokaryons used in the crosses, it is possible to select for certain traits and so increase the chance of producing strains that are able to tolerate a certain set of edaphic conditions. This can be particularly useful for fungi that are to be used for forestry or site remediation work.

## Initial propagation of inocula

Whatever the subsequent method that is used to multiply the culture, the first step is to prepare a culture in liquid medium. To do this, non-solidified sterile medium is transferred to flasks of between 100 ml and 1000 ml. These are inoculated with small blocks of mycelium cut from a selected culture growing in a Petri dish. The flasks are then closed and placed on an automatic shaker to improve aeration and enhance the growth of the fungus.

## Multiplication of the inocula using a solid support

Certain fungi grow slowly in liquid culture, and the main phase of inoculum production is best done in a peat and vermiculite-based solid medium that has been moistened with a suitable nutrient solution. The substrates and the nutrient solution are mixed together and packed into heat resistant sacs and sterilized in an autoclave. After cooling, the substrate in the sacs is inoculated with a small amount of the liquid culture obtained by the technique briefly outlined above. The sacs are then incubated at 25 °C for several weeks to allow the mycelium to colonize all of the available substrate. To use this inoculum to inoculate plants, the substrate must first be rinsed to eliminate any traces of sugar that may favour the growth of opportunistic or pathogenic fungi. The inoculum is then mixed with up to 19 parts of the potting compost typically used to produce the plants to be inoculated. The dilution rate will vary with the vigour of the fungus being used. This method is generally best suited to the inoculation of a small number of plants.

### Multiplication in bioreactor

For the inoculation of thousands or even hundreds of thousands of plants, the mycelium must be produced in a bioreactor. Although this technique is somewhat demanding, it is highly effective and allows the rapid production of large quantities of inoculum. The inoculum produced can be applied in a number of different ways. For example, it can be mixed directly with the potting compost (e.g., in a concrete mixer) prior to sowing, or fragmented and applied to germinating plants using a fertigation system (see above under spore-based inocula).

### Formulation of inocula

To enhance the shelf life of inocula or to provide a product that may be used for applications other than inoculating plants produced under forest tree nursery conditions, it may be necessary to encapsulate the mycelium within a particular support.

Several different techniques have been developed and these range from the use of beads of alginate to entrap small quantities of mycelium or spores that was developed in France, to the inoculum capsules developed by a pharmaceutical company in India. Both these forms are well suited for use in small- or medium-scale plant production systems and they allow the right quantity of inoculum to be placed exactly where it is needed.

CHAPTER 12

# Mycorrhizas in forestry

## Ectomycorrhizal tree species

All woody plant species are mycorrhizal.

Certain gymnosperms (i.e., conifers), such as members of the Taxaceae (e.g., species of *Juniperus*, *Taxus* and *Thuja*), have conserved the ancestral arbuscular mycorrhizal association formed with fungi from the Glomeromycetes. However, over the course of evolution another group, the Pinaceae, gradually abandoned arbuscular mycorrhizal fungi and developed ectomycorrhizal associations. Although some Ascomycetes form ectomycorrhizas, most of the fungi involved are Basidiomycetes.

In the Angiosperms (i.e., flowering plants), the phylogenetic pattern is not so clear. Trees forming ectomycorrhizas are found in the Betulaceae, Corylaceae, Fagaceae, Salicaceae and Tiliaceae. Nearly all the species that do not form ectomycorrhizas, form arbuscular mycorrhizas, and these include members of the Aceraceae, Juglandaceae, Magnoliaceae, Rosaceae and the legumes.

Certain tree species, notably in the Salicaceae and the genus *Eucalyptus*, form both arbuscular mycorrhizas and ectomycorrhizas. Interestingly, the genus *Populus* includes ectomycorrhizal species, arbuscular mycorrhizal species, and species such as *Populus tremuloides* that can carry both types concurrently.

Although members of the Ericaceae typically form ericoid mycorrhizas, certain form ectomycorrhizas, and other form arbutoid mycorrhizas.

## The production of forest tree seedling

The methods used to produce plants destined for reforestation vary according to local customs and the technologies available.

The production of field-grown or bare-rooted plants is still practiced throughout the world. This method is simple and requires less work; however, it has a certain number of associated risks. The principal one being the potential build-up of soil-borne plant pathogens, including nematodes, and pathogenic bacteria and fungi. Because of this, after several years of use, certain nursery soils may become largely unsuitable for plant production, and those plants that are produced may serve as carriers of undesirable micro-organisms. To overcome this problem, the soil may be periodically sterilized using selected biocides, but these normally also eliminate mycorrhizal fungi. Under such circumstances, and depending on the plant species in question, arbuscular mycorrhizal or ectomycorrhizal inocula may be used (see Chapters 8 and 11) to colonize plants with suitable symbiotic fungi.

The production of container-grown plants is of growing importance worldwide, and diverse methods are used. These vary in complexity and may include the use of shading and fertigation systems (Figure 12.1). The substrates used also vary according to the region and resources, and may include forest-collected topsoil, charcoal-based composts, composted plant materials, or peat.

Forest topsoil has the advantage of normally containing mycorrhizal fungal propagules, but it may also be a source of pathogenic organisms. If the soil is sterilized (e.g., using a fungicide or heat) it should be subsequently inoculated with a mycorrhizal fungus. Substrates based on composted materials, such as wood or bark chips, lack mycorrhizal propagules and therefore must also inoculated. In northern and temperate regions, peat-based substrates are typically used for tree seedling production. Industrially produced peat rarely contains undesirable organisms; however, it usually also lacks arbuscular mycorrhizal fungal propagules, which, unlike ectomycorrhizal fungal spores, are not dispersed by the wind. Therefore, for the production of certain species, inoculation with arbuscular mycorrhizal fungi may be necessary. Furthermore, the number of ectomycorrhizal propagules in peat varies considerably and may, in some cases, be absent. Therefore, inoculation with ectomycorrhizal fungi may also be required. Nevertheless, this is generally unnecessary in heavily forested regions, such as the boreal forest, where there is an important natural spore rain. Each fall,

**Figure 12.1** – A modern tree nursery in Tunisia producing container-grown plants. A fertigation system provides water and sufficient levels of soluble fertilizer, and the substrate consists of composted branches of a nitrogen-fixing shrub. The temperatures reached during composting eliminate all ectomycorrhizal fungi propagules and so plants are inoculated early during production, using a spore-based inoculum applied via the fertigation system.

Photo: Mohammed Lamhamedi, MRNF

ectomycorrhizal fungi associated with the roots of container-grown plants, typically produce large quantities of fruitbodies in forest nurseries (Figure 12.2). These are a good source of spores that can be collected and used to produce a spore-based inoculum. In order to ensure the early establishment of ectomycorrhizas, the inoculum can be applied soon after sowing using the fertigation system.

**Figure 12.2** – In boreal forest nurseries, container-grown seedlings often produce a profusion of ectomycorrhizal fungal fruitbodies in the fall. The photo shows *Laccaria bicolor* produced in association with balsam fir.

Photo: Mohammed Lamhamedi, MRNF

Even in nurseries subjected to a heavy natural spore rain, the formation of ectomycorrhizas can be inhibited by unsuitable growing conditions. The tendency for plant producers to focus on the rapid production of large plants that are capable of avoiding natural competition, results in the application of important amounts of nitrogen fertilizer. This prevents the formation of mycorrhizas and leads to inadequate root development, which can limit survival and establishment of the seedling following outplanting.

## Reforestation under arid conditions

Most of the species found in arid and semi-arid climates are associated with arbuscular mycorrhizal fungi. Studies have shown that certain species within the Glomeromycetes are particularly well adapted to these climates, and that selected strains can enhanced the drought tolerance of inoculated host plants. Indeed, they appear to be essential for the survival of twisted acacia, a woody species from the Sahara.

Few species of ectomycorrhizal fungi are capable of surviving in hot arid environments. However, species of *Pisolithus* are an exception and can tolerate temperatures as high as 38 °C. These fungi allow the implantation of trees (e.g., pines and eucalyptus) that are unable to survive if they are not mycorrhizal, or if they are colonized by a non-adapted fungus. The possibility of producing hybrid ectomycorrhizal fungi under controlled laboratory conditions, allows the breeding of fungi that are particularly adapted to enhance the drought resistance of their host. The successful use of such hybrids in plant production relies on the creation and application of suitable mycelium-based inocula.

## Reforestation in calcareous soil

Just as there are particular difficulties associated with the production of plants in areas dominated by calcareous soil, reforestation in these regions also poses a number of problems. For example, the Aleppo pine grows on soils of pH 7 and above. Iron in these soils is not freely available for plant uptake; however, this mineral is essential for the formation of chlorophyll. Plants produced in nurseries that irrigate with hard calcium-rich water and use composts (e.g., composted chipped woody debris) that lack alkaline-tolerant ectomycorrhizal fungi, rapidly become pale green (i.e., chlorotic) and stop growing. However, following inoculation with a suitable ectomycorrhizal fungus and establishment of the symbiosis, the plants resume chlorophyll production and growth (Figure 12.3).

**Figure 12.3** – Inoculation with a suitable ectomycorrhizal fungus corrects iron deficiencies in an alkaline substrate.
Photo: Mohammed Lamhamedi, MRNF

For all forms of reforestation work, plants that are heavily mycorrhizal should be favoured over those that are poorly or non-mycorrhizal.

## The introduction of exotic species

Most of the ectomycorrhizal forming tree species are capable of forming symbioses with most ectomycorrhizal fungal species. However, occasionally the successful establishment of a tree species outside its natural distribution range requires the use of a given species of ectomycorrhizal fungus, or even a given strain. For example, the introduction of Douglas fir from the Pacific Coast of Canada to the Massif Central in France was successful because of the introduction of a fungus from its native region.

## Mycosylviculture and forest management

Chapter 7 highlighted how mycosylvicultural techniques, and controlled management of forests and plantations, can increase the harvest of edible ectomycorrhizal mushrooms for the commercial food market. It is also possible to establish new colonies of Chanterelles and King Boletus by distributing spores of these species in suitable woods and forests.

CHAPTER 13

# Mycorrhizas and the environment

Many different human activities may threaten the biodiversity of mycorrhizal fungi. Among these are agriculture, forestry, large-scale site work, and industrial pollution.

## The impact of agricultural practices

Agriculture, in particular large-scale monocultures, has had a long running and profound effect on fungal biodiversity. For example, less than 100 years of intensive farming in the St. Lawrence River valley (Québec, Canada) has led to a 50% reduction in the biodiversity of arbuscular mycorrhizal fungi at many site. Prior to their use for agriculture, deciduous broad-leafed forests containing over 100 species of arbuscular mycorrhizal plants covered these sites, and their soils contained earthworms and a wide variety of mycorrhizal fungi. Traditional farming methods, using a period of fallow and regular applications of organic matter, favour good populations of earthworms and help maintain a high diversity of arbuscular mycorrhizal fungi. However, this farming approach is rare. In situations where large-scale monoculture systems are used to produce crops such as maize, a no-till approach may help to reduce mineral nutrient leaching, and enhance earthworm and arbuscular mycorrhizal fungal populations.

## The sequestration of carbon in the soil

The mycelial biomass of mycorrhizal fungi comprises an important component of the total soil carbon pool. In a typical coniferous forest in the boreal region, more than 50% of the $CO_2$ released from the soil comes from mycorrhizal fungal respiration.

It has been estimated that one third of the world's soil-sequestered carbon is held as glomalin, a glycoprotein secreted by arbuscular mycorrhizal fungi that is highly resistant to decomposition. Prior to this finding, humic acids were widely considered to be the major component of the organic matter pool in agricultural soils; however, it now seems that glomalin is equally important.

## Legislation on biocides

In most countries, government agencies and regulatory bodies set laws and regulations controlling the use of pesticides, and of mycorrhizal fungal inocula. When it comes to obtaining permission to commercialize an inoculum, the set of criteria that are required are often particularly difficult to meet. However, the same regulatory bodies do not require producers of artificial biocides to show that their products are harmless to arbuscular mycorrhizal fungi. However, such tests would be extremely simple and rapid to conduct using *in vitro* cultures of arbuscular mycorrhizal fungi on root-organ cultures in compartmentalized Petri dishes. Ideally, a vigorous evaluation should be obligatory and all secondary effects on mycorrhizal fungi should be indicated on the product label.

## Site restoration

Sites that have been affected by large-scale construction work or exploitation fall into two categories, those that are toxic and those that are not. Non-toxic sites may include sites negatively impacted during road building or the construction of hydroelectric dams, or certain mining activities that produce spoil that lacks the physical properties and nutrients necessary for the establishment and growth of plants. Excavation of substrates such as gravel, sand and silt, requires the removal of the surface organic layer of the soil. Although the latter is often retained for site restoration work, this is often not sufficient to cover the whole of the affected area. Generally speaking, the materials left after construction or exploitation will allow the reestablishment of vegetation if the substrate comprises a range of particle sizes (e.g., silt, and fine and coarse sand) and if the water table is not too deep. The substrates that are the most difficult to treat are those characterized by a uniform particle size (e.g., just fine sand) with a deep water table.

These substrates invariable lack mycorrhizal fungal propagules, and organic matter and a suitable mycorrhizal inoculum are usually required to ensure their rapid revegetation.

## The revegetation of toxic sites

The residues generated by industrial activities are often toxic (e.g., certain mining waste), and the sites on which they are stocked become unfavourable for the growth of plants. The cause of toxicity varies widely depending on the industry in question, and ranges from minerals (e.g., heavy metals) to organic molecules (e.g., by-products of the oil industry). The solutions to these problems are rarely simple, but there are a number of situations where arbuscular mycorrhizal, ectomycorrhizal and ericoid mycorrhizal fungi offer valuable assistance, and a wide variety of techniques using these fungi are being developed all over the world.

## Soil detoxification

In certain situations and with certain pollutants (e.g., arsenic), mycorrhizal fungi can be used to absorb the toxic substance in question. This is then concentrated in the aerial tissues of the host plants used, which can be regularly harvested and burnt in appropriate facilities. Once this is done, the ash containing the concentrated pollutant, can be disposed of correctly. This technique is only effective if the toxic substance is question is contained within the rooting zone of the plant being used, or if the material to be treated can be spread in a thin layer that is accessible to the roots.

In the boreal forest, the roots of conifers (e.g., pines, fir and spruce) and ericaceous shrubs typically grow in highly acidic soils (pH 3, see Chapter 5). Under such conditions, aluminium, iron and manganese reach concentrations that, in the absence of mycorrhizal fungi, are toxic for plant roots. Therefore, the mycorrhizal fungi present are able to protect their hosts from high concentrations of certain heavy metals. These fungi can be used, not to detoxify soil, but to establish plants on otherwise toxic sites.

## Reducing the risk of algal blooms

It is now clearly understood that the leaching of fertilizers from land bordering lakes can lead to the proliferation of toxic blue-green algae (i.e., cyanobacteria). These algal blooms threaten the health of lake ecosystems, and that of local residents. It is now widely recommended that lakeside property owners establish, and maintain, bands of herbaceous plants and/or shrubs on land adjacent to lake shorelines. This is to reduce the amount of nutrients (above all phosphorous) that seep into the lake. In such locations, the use of fertilizer and even composts must be avoided. An ecologically sound solution for the rapid establishment of vegetation on land adjoining lakeshores is to use mycorrhizal plants. An increasing number of municipalities are now recommending this practice. This is because mycorrhizal fungi not only help the establishment of plants, they also enhance the capture and uptake of surplus phosphorus from the soil.

# Conclusion

It may seem rather audacious to talk of a new green revolution, but if we think carefully about the advantages of incorporating mycorrhizal fungi into plant production systems, it is probably one of the most sensible and promising directions for mankind to take.

The green revolution that occurred during the second half of the 20th century, saw new technologies, such as the use of symbiotic nitrogen fixation and the genetic improvement of plants, integrated into agriculture. The use of nitrogen-fixing symbioses represents an important saving in terms of the energy costs incurred by the transformation of nitrogen from the air into nitrogen fertilizer. Today, symbiotic nitrogen fixation plays an important role in the sustainable production of plants. In the future, its use is likely to undergo further developments and improvements, especially in terms of the efficiency of the associations, and the use of inocula. However, it seems likely that its contribution to agriculture will progress at a slower rate than it has done in the past. For many years, molecular biologists have alluded to the possible development of nitrogen fixing wheat, but this seems to be far from imminent.

While nitrogen availability affects the growth and yield of crops, phosphorus, present in the form of phosphate, is most often the limiting factor. This is particularly true in tropical soils, which have benefited from little, or no, mineral fertilization. In contrast to energy, which can be obtained from a diverse range of sources (e.g., oil, biomass, wind, solar, and nuclear), and nitrogen that we can obtain in apparently inexhaustible quantities from the air, phosphorus seems to be a resource that is non-renewable, and for which there is no substitute. Because phosphorus is a constituent of all living organisms, and because there is no alternative, reduced availability of this element will be a limiting factor for mankind.

In light of this, present agricultural practices must be re-examined. Phosphate must be used wisely and the amounts employed under intensive agricultural systems must be reduced. In addition to the fact that global phosphate reserves are exhaustible, we should also consider the extraction, international transportation, transformation, fertilizer application, and environmental costs incurred by its use. Leached phosphate is a powerful environmental pollutant, and is considered to play a key role in the eutrophication of nitrate-rich waters. It will be important to develop practices that ensure the conservation of phosphate at all stages of its use, and allow crop plants to produce maximum yields under minimum phosphate fertilizer application regimes. While western countries tend to use phosphate fertilizers abusively, India succeeds in producing enough food for its population using very little additional phosphate, even though its soils are relatively poor in this essential element.

One means of economising non-renewable phosphate, is to give greater consideration to the role that mycorrhizal fungi play in the nutrition of plants, and in the functioning of natural and agricultural ecosystems. In the past, phosphate fertilizer application levels were aimed at increasing the soil phosphate reserve to a level beyond that required by the crop. By contrast, modern application schedules are increasingly aimed at managing the short-term needs of the crop and maintaining a basic level of soil fertility. In certain agricultural soils, the prescribed fertilizer application levels could be further reduced if we assure that the soil has an optimal mycorrhizal fungal flora. This supposes that, in the future, the analysis of the quantity and quality of the soil mycorrhizal flora, will becomes standard practice and will be done as routinely as the analysis of soil pH or plant-available nutrients. It also supposes that during pesticide trials, the potentially negative effect on mycorrhizal fungi will be taken into consideration. The negative impact of soil tillage on mycorrhizal fungi should also be taken into account, and direct sowing encouraged. Plant breeders should also make sure that genetic selection is done in an optic that ensures that mycorrhizal fungi, and other soil organisms implicated in the conservation and uptake of phosphorus from natural soil reserves, can play a maximal role in the nutrition of new cultivars.

From a completely different perspective, we also need to consider the other important roles of arbuscular mycorrhizal fungi in the life of plants in natural and cultivated systems. For example, the ability of these fungi to increase the drought stress resistance of crops is completely underestimated. A greater number of large-scale field studies need to be conduct to highlight this phenomenon, especially given that increased drought resistance for as little as one week, can have an important impact of yield, and not only in countries typically affected by drought. For the same reason, buyers of plants sold for outplanting should insist that the plants they purchase are mycorrhizal.

Moreover, scientific studies have demonstrated that arbuscular mycorrhizal fungi can prevent or reduce infections by pathogenic fungi, which could reduce the need for fungicides. Once again, we urgently need to conduct large-scale long-term studies to show this potential in intensive agriculture.

Furthermore, something that has recently come to light is the fascinating effect of arbuscular mycorrhizal fungi on the biochemistry of their hosts. In certain plants, the chemical modifications that they induce, prevents grazing by insect pests. Even more astonishing, is the fact that the flavour of onions and herbs is enhanced by the presence of mycorrhizal fungi in their roots. This leads to the possibility of research and development aimed at improving the taste and nutritional quality of our food. Perhaps these fungi are also responsible for the quality of different wines, cheeses and other food items that is currently linked to an effect of *terroir*.

Under current forestry practices, forests are primarily considered as a source of timber, and this is reflected in the management techniques used. However, in the light of new environmental preoccupations, this approach is being increasingly questioned, and novel ecosystem-based management techniques, which take into consideration the different components of the forest, are being proposed. This new approach encourages the conservation and valorisation of all forest resources, including edible forest mushrooms. Increasingly, the economic value of edible fungi harvested from a given stand, greatly exceed that of the timber it contains. If we consider the question of biodiversity, an important number of fungal species are only associated with old-growth forests; these, and the multitude of other organisms relying on such habitats, need particular protection.

Not only should forest management practices take into account the mycorrhizal fungal flora, they should also give greater consideration to the possibility of mycosylviculture techniques that maintain, or even increase, the production of edible forest mushrooms. If commercial harvesting of fungi can be integrated into plantation management, this could encourage practices that favour fungal fruitbody production and, in the long-term, increase timber yield and quality. Moreover, the sale of fungi could be used to offset the management costs involved.

Finally, progressively more restoration and revegetation projects conducted on polluted sites and mine spoil are using root symbioses, including mycorrhizas. Furthermore, in urban areas, trees planted in parks and gardens, and those used to line roads, which are surrounded by concrete and tarmac, have been shown to benefit from controlled mycorrhization.

For a long time, the widespread use of mycorrhizas around the world has been limited. However, the recent development of techniques allowing the large-scale commercial production of mycorrhizal inoculum in a range of different forms means that it is now possible to envisage a new green revolution. If adopted, this change will make horticultural, agricultural, forestry and restoration practices more cost-effective, sustainable and environmentally friendly. Nevertheless, such ameliorations can only be achieved if each region and country invests heavily in research and development to permit this biotechnology to be adapted to suit their specific environmental conditions, cultivated plants and customs.

It will also be necessary for bodies such as the Food and Agriculture Organization of the United Nations (FAO) to fully understand the potential of the use of mycorrhizas and to become the protagonists, as they did so effectively with the use of symbiotic nitrogen fixation bacteria at the end of the last century.

# Glossary

**Amensalism**: an interaction between two species in which one inhibits the development of the other by depriving it of an essential resource – for example, a tree that limits the growth of plants in the understorey by depriving them of sunlight.

**Antibiosis**: an interaction between two species, where one makes the environment intolerable for the other – for example, a fungus that produces an antibiotic that eliminates bacteria.

**Arbutoid mycorrhizas**: similar to the ectendomycorrhizas of pines, but found in the Ericaceae. They are comprised of a fungal sheath that surrounds the root, an intercellular network of hyphae, and intracellular penetrations.

**Autotrophs:** organisms that produce organic material using the sun's energy (i.e., photoautotrophic plants) or by using chemical energy (i.e., chemoautotrophic bacteria).

**Axenic**: often used to describe an *in vitro* culture containing one organism.

**Axenic culture**: the literal translation means without strangers; and is used to describe a pure culture of a single organism.

**Clay-humus combinations**: In addition to sand, alluvial soils contain finer silt and clay particles. Clay particles are amphoteric, that is, they have a negative and positive pole that can serve as attachment points for labile mineral elements such as potassium ($K^+$). This process prevents elements from being leached from the soil; however, they remain available for uptake by fungi and plant roots. Humus has similar properties to clay particles. In alluvial soils, which are typically rich in arbuscular mycorrhizal fungi, earthworm activity results in clays and humus particles being associated, forming clay-humus combinations. This complex plays an important role in the formation of soil aggregates and thus enhances soil fertility. In podzols, clay particles are often absent, which implies that the mobile elements only come from soil humus.

**Coenocyte**: a cell with multiple nuclei united in a common cytoplasm. The amoeba and arbuscular mycorrhizal fungi are examples of coenocytic organisms.

**Commensalism**: an association between two species where one supplies a resource to the other without receiving any advantage – for example, a hermit crab living in the shell of a mollusc that is already dead.

**Competition**: an interaction between species or organisms for access to resources within the environment – for example, the competition between two plants for mineral nutrients within the soil.

**Cyclosis**: the movement of cytoplasm in a cell. This movement can be circular, as in the epidermal cells of onions, or bi-directional, as in the hyphae of arbuscular mycorrhizal fungi.

**Eukaryotes**: all those organisms in which the cells group the majority of their DNA in a membrane-surrounded nucleus – for example, plants, fungi and animals. During reproduction the cells divide by mitosis.

**Herbivory**: the action of consuming living plant leaves.

**Heterotrophs**: organisms that obtain their energy and carbon from living (i.e., parasites, predators and symbionts) or dead (i.e., saprophytes) organisms.

**Humus**: soil organic matter that plays a key role in soil fertility – there are three main types: mull, moder and mor.

**Inoculum**: in the case of mycorrhizal fungi, it refers to the material used to inoculate a substrate and colonise a host plant. The preparation of inocula is an indispensable stage in the colonization of plants under certain production systems.

**Mixotrophy**: those organisms that obtain part of their carbon and energy from a living autotroph, and the other part by means of saprophytic activity.

**Moder**: this is a type of humus that is intermediate between mor and mull, and that evolves under the influence of the soil microfauna and microflora. However, because of an absence of earthworms due to low soil pH, it is not incorporated into the mineral soil. Under certain circumstances, insect larvae may bring small amounts of humus into the mineral soil, which gives the leached sands a speckled appearance.

**Mor**: this is a type of humus that evolves very slowly. In addition to the absence of earthworms, the phenols and tannins it contains slow bacterial activity to the benefit of fungi. These, in turn, generate the acidity that reduces the pH of the soil. The upper part, which is known as the fermentation horizon, contains recognizable plant structures. By contrast, in the lower part, which is known as the humification horizon, these structures are no longer recognizable. No incorporation of organic matter into the underlying mineral horizons occurs, except for that organic matter that is dissolved by acids and leached towards deeper soil horizons, where the higher pH provokes their precipitation. In the boreal forest, this type of humus is typical of ectomycorrhizal-dominated podzols.

**Moraine**: coarse rocky debris left by glaciers.

**Mull**: this is a type of humus that has evolved under the influence of the soil mesofauna (e.g., earthworms), microfauna (e.g., mites and protozoa), and microflora (e.g., bacteria, actinomycetes and fungi). Plant structures are not recognizable and the humus is incorporated deep into the mineral soil. This type of humus is typical of alluvial brunisols, which are dominated by arbuscular mycorrhizal fungi and highly suitable for agriculture.

**Parasitism**: the interaction between two species, where one obtains its nutrients from the other by partially or wholly destroying the tissues of its host – for example the honey fungus, which penetrates a tree and destroys the cambium.

**Parent material**: the underlying, often coarser, mineral material on which finer particles are deposited to form the soil layer.

**Peat**: organic matter produced as a result of a breakdown of the process that leads to the formation of mor humus. Peat dominated substrates are typically characterized by the presence of bryophytes and members of the Ericaceae.

**Phagocytosis**: a process by which certain organisms ingest and digest particles of organic matter or microscopic organisms directly in their cytoplasm – for example, white blood cells that engulf bacteria.

**Predation**: an interaction between two species, where one eliminates the other by feeding on it – for example, a snake and a mouse.

**Primordium**: the structure that, given the right suite of conditions, may develop into a fungal fruitbody.

**Prokaryotes**: organisms in which the DNA is not within a membrane-bound nucleus – for example, bacteria and actinomycetes. During reproduction, the cells divide by amitosis, which involves a simple separation of the DNA strands.

**Proteoid root**: a plant root that is comprised of a dense group of short lateral roots that are not mycorrhizal. The increased surface area of the root is thought to improve plant nutrition.

**Symbiosis**: a term created by Anton de Bary in 1879 to describe all forms of interactions between two organisms, including parasitism, predation and mutualism. Lynn Margolis forwarded the idea that this term should be restrained to those associations that are mutually beneficial for both organisms. In is strictest sense, in an obligate symbiosis the symbiont cannot live without its host and vice versa.

# For further information


Allen, M.J., 1992. *Mycorrhizal functioning*. Chapman & Hall, New York, London, 534 p.

Ahmadjian, V., Paracer, S., 1986. *Symbiosis: an introduction to biological associations*. Clark University, University Press of New England, 212 p.

Deacon, J., 2006. *Fungal biology*. Blackwell Publishing, Malden, M.A: 371 p.

Declerck, S., Strullu, D.G., Fortin, J.A., 2005. In vitro *culture of mycorrhizas*. Springer-Verlag, Berlin, 388 p.

Fortin, J.A., Bécard, G., Declerck, S., Dalpé, Y., St-Arnaud, M., Coughlan, A.P., Piché, Y., 2002. «Arbuscular Mycorrhiza on root-organ cultures», *Canadian Journal of Botany*. 80 : 1-20.

Fortin, J.A., Charest, C., Piché, Y., 1995. *La symbiose mycorhizienne : état des connaissances*. Éditions Orbis, Frelighsburg, Québec, Canada, 195 p.

Hamel, C., Plenchette, C., 2007. *Mycorrhizae in Crop Production*.The Haworth Press, Inc.10 Alice Street, Binghamton, NY 13904-1580, 319 p.

Khasa, D.P., Piché, Y., Coughlan, A.P., 2008. *Advances in mycorrhizal science and technology*. NRC Press, Ottawa, 14 chapitres (in press).

Margulis, L., Fester, R., 1991. *Symbiosis as a Source of Evolutionary Innovation*. The MIT Press, Cambridge, Massachusetts, London, England, 454 p.

Margulis, L., 1993. *Symbiosis in Cell Evolution*. W.-H. Freeman and Co, New York, 452 p.

Pfleger, F.L., Linderman, R.G., 1994. *Mycorrhizae and plant health*. APS Press, The American phytopathological society, St. Paul, Minnesota, 344 p.

Peterson, L.R., Massicotte, H.B., Melville, L.W., 2004. *Mycorrhizae: anatomy and cell biology*. NRC Press, Ottawa, 173 p.

Plenchette, C., Clermont-Dauphin, C., Meynard, J.M., Fortin, J.A., 2005. «Management of AM fungi in cropping systems». *Canadian Journal of Plant Science*. 85: 31-40.

Rasmunsen, H.N., 1995. *Terrestrial orchids: from seed to mycotrophic plant*. Cambridge University Press, Cambridge, United Kingdom, 444 p.

Read, D.J., Perez,-Moreno, J., 2003. «Mycorrhiza and nutrient cycling in ecosystems – a journey towards relevance? Research Review». *New Phytologist* 157: 475-492.

Read, D.J., Leake, J.R., Perez-Moreno, J., 2004. «Mycorrhizal fungi as drivers of ecosystem processes in heathland and boreal forest biomes». *Canadian Journal of Botany* 82: 1243-1263.

Schliemann, W., Ammer, C., Strack, D. 2008. «Metabolite pofiling of mycorrhizal roots of Medicago truncatula». *Phytochemistry* 69 :112-146.

Seckbach, J., 2002. *Symbiosis: mechanisms and model systems*. Kluwer Academic Publishers, Dordrecht.

Shennan, C., 2008. *Biotic interactions, ecological knowledge and agriculture*. Philosophical transactions of the Royal Society B-biological sciences, 363:717-739.

Sieverding, E., 1991. *Vesicular-Arbuscular Mycorrhiza Management in Tropical Agrosystems*. Technical Cooperation; Eschborn., Germany.

Smith, S.E., Read, D.J., 2008. *Mycorrhizal symbiosis*, Second edition, Academic Press, 769 p.

Strullu, D.G., Garbaye, J., Perrin, R., Plenchette, C., 1991. *Les mycorhizes des arbres et des plantes cultivées*. Technique et Documentation, Lavoisier, Paris, 250 p.

Vega, F.E., Blackwell, M., 2005. *Insect-fungal associations: ecology and evolution*. Oxford University Press, 333 p.

**Researching sustainable systems**. Proceedings of the First Scientific Conference of the International Society of Organic Agriculture Research (ISOFAR), held in Cooperation with the International Federation of Organic Agriculture Movements (IFOAM) and the National Association for Sustainable Agriculture, Australia (NASAA), Adelaide Convention Centre, Adelaide, South Australia, 21-23 September, 2005.

**Cultivating the future based on science**. Volume 1: Organic Crop Production. Proceedings of the Second Scientific Conference of the International Society of Organic Agriculture Research (ISOFAR), held at the 16th IFOAM Organic World Conference in Cooperation with the International Federation of Organic Agriculture Movements (IFOAM) and the Consorzio ModenaBio in Modena, Italy, 18-20 June, 2008.

**Cultivating the future based on science.** Volume 2: Livestock, socio-economy and cross disciplinary research in organic agriculture. Proceedings of the Second Scientific Conference of the International Society of Organic Agriculture Research (ISOFAR), held at the 16th IFOAM Organic World Conference in Cooperation with the International Federation of Organic Agriculture Movements (IFOAM) and the Consorzio ModenaBio in Modena, Italy, 18-20 June, 2008.

## Internet

Collections of arbuscular mycorrhizal fungi
http://invam.caf.wvu.edu/cultures/cultsearch.htm

Company specialized in mycology, Toronto
http://www.sporometrics.com/index.html

Exhaustion of phosphate reserves
http://www.energybulletin.net/33164.html

Global Phosphorus Research Initiative (GPRI)
http://www.phosphorusfutures.net

Glomeromycota in vitro collection (GINCO)
http://emma.agro.ucl.ac.be/ginco-bel/index_fr.php

International data bank: National Center for Biotechnology Information
http://www.ncbi.nlm.nih.gov/

International Mycorrhiza Society
http://www.mycorrhizas.org/

International Symbiosis Society
http://people.bu.edu/iss/

Mycological collections throughout the world
http://mycology.cornell.edu/fcollect.html

Mycological Internet site, Myco Album CD (Dr George Barron)
http://www.uoguelph.ca/~gbarron/

Mycological Internet site, Mycologue Publications (Dr Bryce Kendrick)
http://www.mycolog.com/

MycoQuebec
http://www.mycoQuebec.com

Non-woody forest products
http://www.google.com/search?hl=fr&q=produits+forestiers+non+ligneux+17&btnG=Recherche+Google&lr=

### Commercial *in vitro*-produced inocula

Premier Tech
http://www.usemyke.com

Teri Industry
http://www.teriin.org/mycorrhiza

### Authors

Yves Piché
http://www.cef-cfr.ca/index.php?n=Membres.YvesPiche

J. André Fortin
http://pages.videotron.com/forstamp/

Christian Plenchette
http://c.plenchette.free.fr

# Index

## S



## T